Teaching Statistics

Darren Macey and Will Hornby

CAMBRIDGE
UNIVERSITY PRESS

University Printing House, Cambridge CB2 8BS, United Kingdom

One Liberty Plaza, 20th Floor, New York, NY 10006, USA

477 Williamstown Road, Port Melbourne, VIC 3207, Australia

314–321, 3rd Floor, Plot 3, Splendor Forum, Jasola District Centre, New Delhi – 110025, India

79 Anson Road, #06-04/06, Singapore 079906

Cambridge University Press is part of the University of Cambridge.

It furthers the University's mission by disseminating knowledge in the pursuit of education, learning and research at the highest international levels of excellence.

www.cambridge.org
Information on this title: www.cambridge.org/9781108406307

© Cambridge University Press 2018

This publication is in copyright. Subject to statutory exception and to the provisions of relevant collective licensing agreements, no reproduction of any part may take place without the written permission of Cambridge University Press.

First published 2018

20 19 18 17 16 15 14 13 12 11 10 9 8 7 6 5 4 3 2 1

Printed in Great Britain by CPI Group (UK) Ltd, Croydon CR0 4YY

A catalogue record for this publication is available from the British Library

ISBN 978-1-108-40630-7 Paperback
ISBN 978-1-108-40631-4 Digital edition

Cambridge University Press has no responsibility for the persistence or accuracy of URLs for external or third-party internet websites referred to in this publication, and does not guarantee that any content on such websites is, or will remain, accurate or appropriate. Information regarding prices, travel timetables, and other factual information given in this work is correct at the time of first printing but Cambridge University Press does not guarantee the accuracy of such information thereafter.

....

NOTICE TO TEACHERS IN THE UK
It is illegal to reproduce any part of this work in material form (including photocopying and electronic storage) except under the following circumstances:
(i) where you are abiding by a licence granted to your school or institution by the Copyright Licensing Agency;
(ii) where no such licence exists, or where you wish to exceed the terms of a licence, and you have gained the written permission of Cambridge University Press;
(iii) where you are allowed to reproduce without permission under the provisions of Chapter 3 of the Copyright, Designs and Patents Act 1988, which covers, for example, the reproduction of short passages within certain types of educational anthology and reproduction for the purposes of setting examination questions.

Contents

Foreword	iv
Introduction to the book	vi
Acknowledgments	xi

Part 1: A vision for statistics in schools — 1

1	What should all students understand and why is it important?	1
2	The statistical cycle	14
3	Exploratory data analysis	32
4	Simulation	48
5	Sampling and variation	58
6	Signals and noise	71
7	Informal inference	81
8	Technology in the classroom	94

Part 2: Activities — 104

	Introduction to the activities	104
9	Activity for Chapter 2	106
10	Activity for Chapter 3	113
11	Activities for Chapter 4	119
12	Activities for Chapter 5	128
13	Activity for Chapter 6	134
14	Activity for Chapter 7	139

Appendix 1: Useful R commands	144
Appendix 2: Some prompts for investigations	154
Index	188

Foreword

There's no getting away from it – for better or worse, we're now surrounded by data. Unless we live totally isolated lives, we are presented or confronted by facts and statistics which are measured, collected, visualised, reported and analysed by many organisations and individuals everywhere, every single day. Of course, if we'd lived a century ago, our access to such information would have been severely limited by many factors – the most obvious of which is technology. Now we can access information at the touch of a button, on the move, at work, at home or at play. One might ask: *In what ways have our lives been changed by this wealth of data?* We certainly have access to more information, but does that necessarily mean we are better informed? Does more accessible data mean we make better decisions?

Teaching Statistics is the second book in the series, the first being the excellent Gage and Spiegelhalter: *Teaching Probability*. Both emphasise the importance of comprehending a world in which we are surrounded by data, and both aim to give the reader an understanding of how we can ensure that the next generation *can* make better decisions. They do this through identifying two big issues.

Part One sets the scene for effective teaching. Firstly, the authors argue that the way that statistics, or data handling, is currently taught in UK schools (and in many other jurisdictions too) largely misses the point of studying the subject. Data is not just about numbers – it's about numbers in context. Each part of the data handling cycle is influenced by, and contributes to, the understanding of that context. Fluency in, for example, calculating the mean, is 'meaningless' unless the mean is situated in a context which allows the student to say something about its importance. Similarly, being asked to represent a set of non-contextualised data in several different ways, without having the information to decide on which kind of diagram would best illustrate it, makes a nonsense of the activity. The authors argue strongly that the vast majority of statistical learning should take place within the statistical cycle, so that students understand that the data tells a story and the way it is gathered, visualised, reported and analysed needs to take that into account.

The second issue that makes this text innovative in helping the next generation to make better decisions is the centrality of technology. The authors accept that digital resources are not in everyday use by all teachers but describe how such software supports sophisticated visualising, simulating and analysis in a way that is not possible without it. Whilst the activities they suggest include those with and without technology, they emphasise their different affordances and encourage and support reluctant teachers to 'have a go'. The appendix illustrating 'R commands' is particularly useful for this.

Foreword

Part Two exemplifies the philosophy of Part One. Each set of activities links to a chapter in Part One and brings the theory in it to life. This is where the reader begins to appreciate the power of statistics, making the mathematics relevant. Activities involving real data sets can be powerful levers for engagement and these activities are well-described and accessible. The last section of the book illustrates a perhaps unexpected source of ideas for the classroom by looking at some recent examination questions which involve prompts for investigative work.

This is a very practical book, with plenty of ideas for teachers to use. However, it is rooted in a serious study of the literature and research evidence of the subject. It is this mixture of the practical and the theoretical that makes the text one that is both useful and authoritative, and I'm delighted that it is associated with Cambridge Mathematics.

Lynne McClure
Director, Cambridge Mathematics
Cambridge, Spring 2018

Cambridge Mathematics is a University of Cambridge initiative which seeks to develop a coherent, transparent, evidence-based vision for mathematics education.

This book meets the following Cambridge Mathematics criteria:

- Demonstrates support for connections between mathematical themes
- Demonstrates support for conceptual development
- Demonstrates support for both conceptual understanding and procedural fluency
- Demonstrates that higher order thinking skills, reasoning and problem solving are appropriate for all ages and attainments
- Identifies appropriate historical and cultural links
- Makes explicit the design principles and the underpinning research and evidence
- Demonstrates the modelling of mathematics without relying on surface explanations

For a full set of the Cambridge Mathematics criteria see cambridgemaths.org

Introduction

Introduction to the book

Every day, citizens around the world are bombarded with statistics, from trivial things like the preferences of 8 out of 10 feline consumers, to the relative certainty of scientific claims about new fundamental particles of matter. Statistics are used to demonstrate the superiority of a product over its competitors, illustrate the arguments of journalists and politicians, form the basis of public policy, and even predict the future. In the modern world statistics are everywhere; the ever-increasing availability of 'big data', the name given to huge repositories of information such as Google searches, shopping habits, astronomical observations and world climate, means that few decisions are taken without some form of data analysis.

While it is generally understood that the study of mathematics is an essential component of modern education, most higher qualifications lead students up a path of ever more complicated algebra, culminating in exploration of quadratics, functions and, at the pinnacle, calculus. But when was the last time you opened a newspaper and read an article in which the reporter's argument was backed up by a rather neat factorisation? Or watched the news and saw a politician deliver a sound bite dripping with differential equations?

These relatively advanced aspects of mathematics are fundamental for anyone going on to further study in mathematics and related disciplines such as the physical sciences, but for much of the population they rapidly sink into a pool of barely remembered things which seem to have been inflicted by a teacher for no discernible reason. There is of course an essential subset of the mathematics curriculum which aims to equip citizens with the skills needed for interacting with the world after leaving formal education. There is also great fascination and beauty to be found in the aspects of mathematics that students are expected to learn but will rarely use directly outside of the classroom. The absence of many of the topics currently taught would impoverish the mathematical experience of many students, but it's harder to argue that this absence would have a material impact on students' ability to function in society later in their lives.

Statistical ideas, on the other hand, could be argued to be the summit of the mathematical mountain for everybody not taking mathematics-related study beyond secondary level. Whatever their chosen career path, whatever their level of educational attainment, it is almost certain that a person will be presented with the outcomes of statistical analysis at some point when making decisions for daily life. It is also likely that the most sophisticated mathematical ideas that the majority of people are exposed to will be based on statistics, either used or abused, in order to affect their opinions. The result of this is that a lack of emphasis given to statistical literacy really does have a material impact on students' ability to function in the modern world.

The study of statistics is advancing rapidly; it is a field that has developed in tandem with the advance of computer science, allowing vast quantities of data to be collected and processed in ways that were inconceivable just a decade ago and leading to entirely new techniques of predictive analysis. Education inevitably fails to keep pace with this change despite the best efforts of many excellent classroom practitioners and advocates. In many areas curricula even fail to reflect some of the fundamental basics of good statistics.

Many of the difficulties encountered when learning about statistics have a basis in language as much as in technical competence. Even the word 'statistics' can be used interchangeably as:

> *statistics – the mathematical discipline related to the collection, analysis and presentation of data*

or

> *statistics – the plural of statistic, a parameter calculated from a data set.*

The language issues just get worse from here, with 'describe what the data tells us about girls in our class' and 'describe what the data from our class tells us about girls' demanding entirely different interpretations. At this point we hesitate to even bring up the difference between a sample distribution and a sampling distribution!

If as educators we wish to prepare students to be able to critically evaluate the statistics that will be thrown carelessly towards them throughout their lives, it is time to consider whether leaving school knowing how to draw by hand a selection of standard diagrams and calculate the mean, median, mode and range of raw data is sufficient or even necessary. If we are preparing students for the summit of the mathematical mountain, we must lead them at least as far as base camp, rather than abandoning them on the local hillside with one crampon and a hand-drawn map of Nepal.

It is encouraging that reforms in education are beginning to focus more on how statistics are used and less on how they are calculated, but statistical literacy, the skillset that allows citizens to critically evaluate information in whatever form it is presented to them, must not be treated as a higher-order skill which is left until too late in the curriculum for most learners to access it. If we develop curricula that treat *calculation* as the foundational skill necessary for all, then we let down those individuals who leave education without high levels of academic achievement.

The approach to statistics in this book is based on the Cambridge Mathematics Framework, which has at its core the interconnectedness of mathematical ideas and the integrity of mathematics as a discipline. The focus is on the experiences that students should have of statistics in order to develop critical thinking and a solid conceptual framework. This is not to say that there is no place for calculation, but the first experience of a statistical concept should not necessarily be learning how to calculate measures related to it. It is often possible, and we argue better, to experience drawing inference from data before formalising the practice through calculation and analysis. This liberation from formal *techniques* allows statistical *concepts* to be introduced much earlier in the curriculum.

This approach to learning mathematics is not new; the Cockcroft report in 1982 made similar assertions with regard to the teaching and learning of mathematical concepts, pointing out that:

'There are certainly some things in mathematics which need to be learned by heart but we do not believe that it should ever be necessary in the teaching of mathematics to commit things to memory without at the same time seeking to develop a proper understanding of the mathematics to which they relate.'

Many reports both before and since have argued similarly. We live in increasingly polarised times and it's easy to be convinced by one of the opposing parties arguing for either rote learning techniques or rich tasks and experiences. We prefer a balanced approach, encouraging students to explore ideas and concepts initially without the limitations of explicit calculation and right/wrong answers, before formalising concepts and engaging in structured practice. It is important to note that the exercises chosen to support the development of fluency in calculation should borrow from the best of the ideas championed by such high-performing jurisdictions as Singapore, Shanghai and Japan, but be equally relevant in many UK classrooms, where each question is an opportunity not just to repeat a learned calculation, but also to develop sophistication in response and explore misconceptions.

To give a straightforward example, the study of histograms can be problematic for students; many concepts are wrapped up within what appears to be a straightforward diagram. Subtleties of continuous data, representation of frequency by area, and uneven class widths mean that this essential representation of data appears at a late stage in many curricula which focus on training students to generate their own histograms by hand.

Introduction to the book

It doesn't have to be like this though. Histograms with equal class width could be introduced alongside other diagrams such as pictograms at an early age, providing students with opportunities to read data from them or to describe the shape of the distribution and what this means in context of the data. Later on in their studies, students could explore histograms using a computer and raw data to investigate, for example, how changing class widths affects the usefulness of the diagram. Finally, the most able students may be asked to draw histograms by hand, but at this point the question must be asked, why? Does anyone benefit from drawing such graphs with a ruler and pencil? A compelling case can be made for the pure mathematician being able to quickly sketch a function by hand on the table top of a university common room, but it's hard to imagine anyone with a presentation to prepare on customer income demographics choosing to draw diagrams by hand rather than making use of their desktop spreadsheet software.

Education must provide a model for how statistics is used in the modern world, where longitudinal data and randomised control trials are as relevant as the cross-sectional data that currently forms the context for most early statistical study, and where most data are in the form of a multivariate data set. This is a far cry from the univariate and bivariate data that many statistics courses begin with.

In this book, probability will be placed at the heart of the process of learning statistics – not as a mathematical application of combinatorics, but as a tool for generating data for analysis and for understanding the outcome of experiments. As with the previous example of histograms, randomness and variability can be experienced early on, through games of chance and simulation, without the need for formal mathematical calculation. This opens up a pathway for students to reach richer analytical conclusions from data. A deeper study of probability will allow students to comment on the reliability of their conclusions by informally comparing data to model distributions as a matter of course, rather than having this be reserved for the limited number of students who explore hypothesis tests in pre-university study.

Statistics is also a powerful social tool providing opportunities to inspire young people to engage with important issues such as climate change, sexual health and community cohesion. While a certain degree of bravery is necessary on the part of teachers, a curriculum for statistics that grasps the opportunity to explore these difficult social contexts, both by exploratory data analysis and through discussing published data and journalistic articles, could enhance students' statistical literacy as well as acting as a vehicle to confront the important scientific and sociological issues of the day.

The goal of this book is to describe the essential content of a modern statistics curriculum and to present this content as part of a coherent subject in which data are used to ask and answer questions and the limitations of those answers are fully understood. The activities in Part 2 of the book will focus on statistical literacy, and exemplify the experience-based approach of Cambridge Mathematics. They provide a starting point for teachers to design schemes of work that have at their core a model of statistics that will prepare students to engage with the world full of data in which we all live.

How to use this book

Each chapter in Part 1 is designed to stand alone and explore a specific aspect of statistical literacy. Chapter 8 describes how to create a resource for a classroom demonstration of the variability of sampling using three different pieces of software available to most teachers. The purpose of this chapter is to provide teachers with an entry point into technological aids that they may not have encountered before.

The activities in Part 2 relate to the theory and practice discussed in the chapters in Part 1. We recommend reading each chapter in Part 1 along with its related classroom activity or activities described in Part 2.

References and other sources

1. Cockcroft, W. (1982). *Mathematics Counts: Report of the Committee of Inquiry into the Teaching of Mathematics in Schools* (London: HMSO).

Acknowledgements

The authors and publishers acknowledge the following sources of copyright material and are grateful for the permissions granted. While every effort has been made, it has not always been possible to identify the sources of all the material used, or to trace all copyright holders. If any omissions are brought to our notice, we will be happy to include the appropriate acknowledgements on reprinting.

Fig 2.4 ELIZABETH PHILLIPS, CONNECTED MATHEMATICS 3 TEACHER GUIDE GRADE 6 DATA ABOUT US: STATISTICS & DATA ANALYSIS COPYRIGHT 2014, 0 Ed., ©2014 reprinted by permission of Pearson Education, Inc., New York, New York.

Fig 2.5 from OCR, GCSE Mathematics, Paper J567/03, June 2015

Table 2.1 Reprinted with permission from Principles and Standards for School Mathematics, copyright 2000 by the National Council of Teachers of Mathematics (NCTM); NCTM does not endorse the content or validity of these alignments.

Fig 3.4 graphs from Tufte, Edward (1983) *The Visual Display of Quantitative Information*, 2nd edn 2001 published by Graphics Press USA, based on Paul McCracken, et al, *Towards Full Employment and Price Stability* (Paris, 1977), p. 106, Organisation for Economic Cooperation and Development, Paris.

Fig 3.5 Trilogy meter by Dan Meth, used with permission.

Fig 3.6 screenshot from www.gapminder.org

Fig 5.2 from 'Scaffolding students' informal inference and argumentation (2006)' used by kind permission of Professor Dani Ben-Zvi.

Fig 7.3 from 'Towards more accessible conceptions of statistical inference' C.J. Wild, M. Pfannkuch, M. Regan, N.J. Horton, in the Journal of the Royal Statistical Society: Series A (Statistics in Society), March 14, 2011.

Fig 14.1 is from *Teaching Probability* by Dr David Spiegelhalter and Dr Jenny Gage (Cambridge University Press, 2016), photo by Dr Jenny Gage.

Appendix 1 Task 5 photos (left) VictorHuang/Getty Images, (right) Tom Cockrem/Getty Images

Appendix 2 from OCR, GCSE Mathematics, 1962/08/2345/2318, 2002;

OCR, GCSE Mathematics, 1962/08/1966/2345/1968/2318/1969/05, 2003;

OCR, GCSE Mathematics, 1962/08/2345/2318/1969/05, 2004;

OCR, GCSE Mathematics, 1962/08/2345/2318/1969/05, 2005;

OCR, GCSE Mathematics, 1962/08/2345/2318/1969/05, 2006;

OCR, GCSE Mathematics, J512/06/B254/B266, 2007

Cover image Michelle Patrick/EyeEm/Getty Images

PART 1 A vision for statistics in schools

Chapter 1

What should all students understand and why is it important?

1.1 Who needs statistics anyway?

When planning a statistics curriculum for all, the result should be dependent on the end users. It is convenient, if reductive, to categorise users of statistics based on the level of integration of statistical concepts into their daily lives. Of course, the careful statistician will recognise that although this is a helpful process for building a model of essential skills, any individual member of the population is unlikely to fall neatly into a single descriptive group.

Expert statisticians	People who work with data routinely and analyse the results directly.
Functional statisticians	People who receive data and statistics second-hand and use them for pre-defined, possibly standardised, tasks.
Occasional (and perhaps unwilling) statisticians	People who encounter analysis based on data and use this to make decisions.

Table 1.1 Users of statistics

Expert statisticians include scientists, academics, engineers and people who might calculate a p-value without a second thought. These people are capable of being utterly fascinating or staggeringly boring at parties depending on the perspective of the observer. They have probably gained many of their statistical skills as part of a degree-level qualification. For this group, much of what they have learnt in school about statistics has been supplanted by sophisticated new techniques supported by powerful software packages. Despite this, examples exist of the use of data leading to unsafe conclusions: many examples are detailed by Dr Ben Goldacre at www.badscience.net and provide an excellent source of stimulating material for interested students.

> ### The *p*-value problem
>
> It is not just 'bad' science that produces potentially dangerous results of course. At the heart of the scientific method is the hypothesis test, a confirmatory test that essentially examines whether the data are unlikely to have happened by chance. If they are unlikely to have happened by chance, then this an indicates that there is something interesting going on.
>
> But how unlikely is 'unlikely'? The standard for publication in most cases is about 5%, so in fact somewhere around 1 in 20 scientific studies are drawing their conclusions from data which just happened to come out looking a bit interesting by chance. This is unlikely for any particular individual study, but there are thousands and thousands of studies published every year.
>
> This phenomenon is known as 'the *p*-value problem', and there is an ongoing debate in the scientific community about the level of confidence there ought to be in a conclusion before publishing it.

For future expert statisticians, a secondary-level course in statistics does not need to introduce them to the sophisticated techniques which will be learned later in the context of their area of study. It should, however, prepare them to be critical when engaging in analysis and experimental design, introducing them to the idea of selecting appropriate techniques rather than reaching for the first tool in their statistical toolbox. Students should gain a sense of the limitations of statistical analysis and the need to allow the wider context to inform any decisions and conclusions.

Functional statisticians are those who work in jobs where spreadsheets appear with sometimes alarming regularity in their email inbox. These people may have little experience beyond what they studied at secondary or college level, but will be expected to make judgements and predictions based on trends, averages and raw data. Many teachers fall into this category since student progress is increasingly judged by recorded data rather than professional insight. Often, teachers and senior leaders are not adequately supported to make the data-based predictions of student outcomes currently required. It is not uncommon in UK schools for teachers to be pressured to justify why an individual student has achieved below a target grade that was based on a third-party model of progress for 'equivalent' students. This sometimes happens even when a class of 30 students or a cohort of 200+ has achieved or exceeded the overall targets predicted by the model. This lack of understanding that a model may be used on a macro scale but is unsafe as a predictor of individual performance leads to considerable stress being inflicted on teachers and students due to a fundamentally incorrect assumption. It would be naïve to assume that this kind of misinterpretation is limited to the teaching profession alone! For functional statisticians, a secondary course in statistics should provide them with knowledge of the limitations of data for inference and the ability to defend themselves against unsafe conclusions.

Occasional statisticians are those who work in jobs where they do not encounter data directly and have little need for it in their home life. These are the people most let down by the traditional approach to learning statistics. Every day they are bombarded with information and data that has been abused and wrangled into TV ads. They read web pages and newspaper articles that tell them their risk of cancer will go up if they drink too much red wine, and down if they eat vast quantities of kale. They vote in elections after watching politicians and journalists quote seemingly contradictory statistics demonstrating both that everybody is better off and that society is disintegrating around them. During their time in education they may have been unwilling to engage with a statistics curriculum which appeared irrelevant to, and disconnected from, their daily lives, so they do not have the skills needed to critically evaluate these statements and arguments. For the occasional statistician, secondary education must provide the skills to recognise when statistics are being used well, and when data are presented badly. It is often said that you can prove anything with statistics, but this is not the case. You can *imply* anything by misuse of statistics, and some would suggest it is a right of everybody completing compulsory education that they leave with the statistical literacy needed to understand and critically evaluate statistics and data-based arguments.

In his book *Thinking, fast and slow* Daniel Kahneman described how our brains work against us, convincing us that what we see and experience directly leads to mental models that we then over-generalise, often giving us an inadequate or skewed perception of the world around us. Statistical literacy is the ideal toolkit for exploring this concept in school so that individuals can critically evaluate their own world-view in the future and use data to challenge their perceptions of the world around them.

A secondary curriculum for statistical literacy should meet all the needs of the occasional statisticians, should be flexible enough to provide the functional statisticians with the skills they will need to make sound judgements, and should lay the conceptual foundations for the expert statisticians in anticipation of more technical study later.

1.2 Towards a new type of statistics curriculum

Many current curricula appear to work from a different assumption: that everybody will be an expert statistician. Students' journey through statistics in school is a trajectory subdivided with this goal in mind.

Because statistics is treated as a mathematical discipline, progression is based on increasingly complex calculations and diagrammatic representations. Only superficial levels of interpretation are necessary and inference is treated as a higher-order skill encountered only during further study. Recent research (Makar & Rubin 2014) into informal inferential reasoning has found that students at primary level are capable of discussing likelihood based on simple data in familiar contexts and, with carefully planned tasks, can make reasoned decisions despite few sophisticated techniques being available to them.

So, what do students need to know? Cobb (Steen 1992) argued that any introductory course should explore the following ideas.

1. The need for data.

2. The importance of data production.

3. The omnipresence of variability.

4. The quantification and explanation of variability.

This provides us with a good starting point for considering the aims of our curriculum, but to decide on the focus and content a little more detail is needed. Building on this work, Garfield (1995) suggested that a college-level initial course in statistical literacy should be based on the following foundations.

1. The idea of variability of data and summary statistics.

2. Normal distributions are useful models though they are seldom perfect fits.

3. The usefulness of sample characteristics (and inference made using these measures) depends critically on how sampling is conducted.

4. A correlation between two variables does not imply cause and effect.

5. Statistics can prove very little conclusively, although it may suggest things, and therefore statistical conclusions should not be blindly accepted.

The most striking thing about many research papers on statistics education is the lack of discussion of the merits of training students to perform standardised statistical calculations and produce formalised graphical representations; yet most curricula rely on this type of learning to form the backbone of their content. So why is this still the case when expert opinion has been pointing to a different approach for several decades, especially when the underlying ideas are accessible to young students if technical expertise is not introduced as a barrier?

There are several things to consider. Firstly, statistics is a relatively new discipline compared to much of school mathematics. While the cutting edge of mathematics is constantly advancing, the bulk of the mathematical content taught in modern secondary classrooms can trace a direct line back to luminaries like Euclid and Pythagoras over 2000 years ago. Statistics on the other hand has developed much more recently, and the fundamental principles have advanced rapidly along with the rise of computing power. Taking the study of probability (a key building block of statistical inference) as an example, while there is historical evidence that the ancient Egyptians enjoyed dice games around 3500 BCE, it was not until Blaise Pascal and Pierre de Fermat began investigating problems of probability in the 1600s that a formal mathematical theory began to develop (Lightner 1991). Modern computing methods allow the creation of probabilistic simulations in the classroom unimaginable just a few decades ago. New technology has transformed the interaction between probability and statistics and allowed powerful techniques of analysis to be applied to the growing discipline of statistical inference from so-called 'big data', the collection of vast databases of stellar observations, purchasing patterns and even search histories.

Modern computing allows the almost instantaneous generation of graphical representations of data, some of which conform to none of the standardised charts and graphs that people have become familiar with during schooling. This means that unless a specific pedagogical value can be attributed to training students to construct by hand bar charts, stem-and-leaf diagrams, and all the other graphs that most school leavers will currently be familiar with, it is hard to justify spending large amounts of time doing so.

Most research into the development of graphing techniques advocates modelling with data (Lehrer 2007) to answer simple questions posed by students, with time spent inventing meta-representations (diSessa 2004) – charts and tables devised by the students that do not conform to any pre-defined method. Over an extended period of time, students discuss and refine their models, eventually moving towards more standard representations appropriate to the type of data they are working with. While this process necessarily begins at primary level as students move from object-based representations (sorting objects into groups and arranging them physically to create 'object graphs') to iconic representations (using drawings and stickers to represent data before moving towards crosses and bars), it should continue into secondary school as students begin to explore new kinds of data such as continuous grouped data and work towards developing the more formal representations against which they will be assessed.

A far more important skill to focus on is selecting *useful* representations when a computer can dumbly generate multiple diagrams without consideration of their appropriateness for the data. On top of this, without the ability to interpret unfamiliar representations in the form of infographics and the like, useful information will be obscured. It is incredibly difficult for school curricula to keep up with advances in the field, and any changes that are made require substantial investment in training for the teaching workforce.

A second reason for the disconnect between expert recommendations and practice is due to the status of statistics as an aspect of mathematics. In one key way statistical literacy is fundamentally different to other areas of applied mathematics, especially at secondary level: it is completely reliant on context and interpretation. The study of statistical literacy can be thought of as the study of numbers in context. It is the context in which data exists that makes sense of statistical results, and ignoring the context reduces the study of statistics to an exercise in following a set of recipes that can be mechanistically repeated without understanding their purpose. Unfortunately, as statistical literacy occupies a place in the mathematics curriculum, the focus of study is on the mathematical characteristics that can be refined to formulae and processes, rather than the more difficult-to-define interpretive aspect that relies on discussion and a broad understanding of the wider context. Defining the study of statistical literacy as a purely mathematical discipline is akin to defining the construction of furniture as the assembly of a flat-pack kit. Most people can create a serviceable cabinet by following instructions and putting together some pre-made pieces with standard tools, but the end product lacks the quality of a handmade item and almost always has to fit in a space whose dimensions are at odds with the size of the finished article.

The third reason is assessment. Education is obsessed with the measurement of progress, and it is convenient to create a hierarchy of ever more complicated calculations on which to base judgement of students' ability to 'do statistics'. This is an additional hangover from the treatment of statistics as purely mathematical objects. It is rare in formal, summative mathematics assessment that the sophistication of a student's response to a question is measured. The most prevalent form of assessment question has a single 'correct' answer with perhaps several possible methods of solution available, all receiving equal credit. With statistics, however, requiring a single correct solution may strip the context and the nuance from the data. This is peculiar to mathematics, as most other subjects in school are much more comfortable with the idea of students answering questions with differing levels of sophistication, all of which gain a reasonable amount of credit. Consequently, while the mechanics of calculating statistics may well best be taught by maths teachers, the interpretative aspect and eventual assessment would arguably be better delivered in subjects where the context is key, such as geography, biology and psychology amongst many possibilities.

1.3 The ability to interpret diagrams

Because of the increasing ease with which data can be presented using computers, students must leave school with sophisticated skills in the interpretation of data presented in diagrammatic forms. Currently curricula tend to focus on a small number of key ideas:

- types of correlation (strong/weak, positive/negative)
- identifying the modal value/class
- identifying outliers
- comparing range / interquartile range
- identifying the median and calculating the mean.

While these are straightforward to assess, they are almost useless as procedural skills when classroom practice meets the messy real world. Take scatter graphs for example: it is rare that the kind of neat and tidy examples of positive or negative correlation found in mathematics classrooms exist in data collected in the real world. Far more likely is that a scatter graph will have regions in which there seems to be a stronger correlation, and regions where the association appears less clear-cut. In these cases, the skill is in deciding whether the lack of clarity is a result of the natural variation in the data distorting a genuine relationship, subpopulations in the data with different strengths of relationship, or a coincidental pattern that appears to show association where none exists. The ability to consider these different possibilities allows individuals to make informed decisions when such data is presented alongside a specific interpretation. In many cases the perspective or bias of the person sharing the data will have an impact on the given interpretation, and it is essential that this is always considered.

1.4 An understanding of variability

Understanding variability is fundamental to understanding what is going on with statistical processes and making good judgements. Variability is present in all stages of statistical investigation and has an impact on the entire process from data collection through to any eventual interpretation. The job of a statistician is to understand the implications of variability in the data and to minimise the aspects of it over which there is a degree of control. It is important to identify the types of variability that can occur in data. Firstly, variability caused by inaccuracy in measurement can be minimised by well-designed methods of data collection but never eliminated entirely. Secondly, some variability is inherent in the object of study as most real-world processes are not deterministic. This aspect of variability provides important information for the statistician, and the relative size of this variability in relation to the context of the data will have a big impact on any conclusions that are drawn. Finally, there is an element of variability caused by sampling from a larger population or populations, which again has an impact on the quality of the inferences that may be drawn. Many classroom activities involve comparing two related data sets, such as heights of girls and heights of boys. Students are encouraged to compare measures of average and spread, but they should also question whether they are dealing with a population, which would allow definitive statements of comparison to be made, or a sample, with an underlying variability that may make conclusions unreliable or demand further investigation.

1.5 Sampling and populations

Statistics can be used as parameters for populations, allowing us to make definitive statements about the populations' composition or to make reliable comparisons between them. Unfortunately, due to the way statistical literacy is often taught, inferences made from samples are treated as if they were solid conclusions drawn from populations. The difference in height of boys and girls in class 4B is treated as a valid indicator of the difference in height of all boys and girls in a population with little consideration of the error in the calculated statistics. This is particularly evident in assessment questions where students are asked to make specific comments on differences and similarities in data from diagrams such as boxplots, but are given no opportunity to gain credit by discussing whether the differences are statistically significant based on the sample presented.

It can be a tricky concept as the distinction between population and sample can be related to the language used in the question. 'Does the data show that girls in class 4B are generally taller than boys in class 4B?' treats the data for the class as an entire population. 'Using your data for class 4B, are girls generally taller than boys?' treats the data as a sample, meaning that any conclusions must be treated differently.

Sampling is a fundamental aspect of statistical analysis, with significant consequences for the decisions that can be made and the validity of any conclusions that are drawn.

It is not enough to learn that 'a random sample should be taken to ensure that the data are representative'. It would surprise many people to know that in a lot of cases, a random sample may result in a data set that is significantly less representative. Imagine for example that we want to test the hypothesis that 'students enjoy stats lessons'. Our population is a local secondary school in which students of different ages and abilities are taught by different teachers. The likelihood is that each student's experience of these lessons will be dependent on both their age (affecting the content being taught) and their teacher. In a truly random sample, where every student has an equal likelihood of inclusion, it is not inconceivable that the vagaries of chance may result in proportionally more students being selected from a single class. If this is the class of a teacher whose inspirational approach to teaching, inspired by Robin Williams, has students standing on their desks and passionately reciting the formula for calculating standard deviation before undertaking a spot of extra-curricular data analysis, an overly positive assessment of the popularity of the subject may result. It is essential that experimental design takes into account variability in sampling to eliminate bias without introducing constraints that could obscure signals in the data.

A thorough understanding of sampling methods would enable students to critique the methodology behind statistical conclusions. The goal should be for students to be able to look behind the data presented and consider whether the conclusion is safe. Many advertisements now give some details of the sample used to justify any claims made, and this can be illuminating to those prepared to look critically at the sample and ask questions such as: how many people were involved? What proportion was positive? Were the participants independent of the product?

On top of this, if students left school with some understanding of the power and importance of double blind trials and randomised control trials, it would give them the great advantage of being able to recognise when poor-quality studies are being presented as definitive research and to assess the validity of the conclusions accordingly.

1.6 Distributions and their shapes

Statistics is often perceived as a method of taking a large set of data and, through a process of mathematical alchemy, turning it into a single, easy to use, numerical value. Currently students are comfortable making direct comparisons between the median values for two data sets and will receive credit in assessments for doing so. They may also make some brief comment on the range or interquartile range shown on a diagram. This process of 'reading the data' (Curcio 1987) is the least sophisticated analysis technique and does not consider the global perspective (the complete data set). Students need to be able to go beyond the values they read from the graph and consider the implications of the shape of the data. How does the spread of the data affect the validity of the median value? How does the scale of the diagram affect the perception of the total spread? Does the data set have a large skew or long tails? The implications of the answers to these questions must be considered when deciding if the calculated statistics are meaningful. If the data is a sample from a larger population, the effect of sampling variation must be considered too.

1.7 Correlation and causation

Data in the modern world are a genuine commodity, with companies trading massive data sets collected from their customers using tools such as mobile phone apps and web pages. It is not uncommon when buying goods or services on the internet to be forced to decipher the language of pre-sale tick boxes to communicate that as a customer 'Yes, I agree to the terms and conditions' and 'No, I would not be happy for my details to be passed on to carefully selected partners'. This data is often used to identify correlations in order to model and predict consumer habits to target advertising in the future.

The technique of 'exploratory data analysis' is a relatively new one that has developed alongside the rise of large, multivariate data sets. Where statistical analysis was once based on creating a hypothesis and then collecting the necessary data to demonstrate if the hypothesis was true or not, exploratory techniques rely on the use of technology to analyse data for interesting results and then seeking to explain what the underlying reason for the results might be.

The *Gapminder world* website (www.gapminder.org) contains multivariable demographic data for countries that can be plotted using bivariate representations, allowing associations to be spotted. The ability to easily identify associations without first hypothesising that one exists makes it more important than ever that students are aware of the old mantra that 'correlation does not imply causation'. A strong relationship identified between two variables may in fact be the result of a hidden variable that is driving both, or even reflect an underlying selection bias. Some thought must also be given to how outliers and sample/population size affect the conclusions drawn. It is straightforward to find a correlation coefficient using spreadsheet software, but such statistics can be strongly influenced by outliers or subpopulations, and it is important that both visual and numerical techniques are used to gain a clear idea of what is going on.

1.8 What's changing?

Many of the concepts detailed so far go beyond what is currently taught in classrooms in preparation for end-of-course assessments. This is largely because the current style of assessment does not reflect the experience of statistics that people will have outside the school system and focuses narrowly on the mathematical aspects of the discipline while neglecting the contextual and philosophical aspects.

Reports such as the *Guidance for assessment and instruction in statistics education* (Franklin 2007) and some curricula are recommending a more holistic approach to statistical literacy. They suggest the inclusion of investigation, analysis and statistical literacy with the intention of better preparing students to engage with data beyond compulsory schooling. In this book we seek to reflect much of the approach promoted by the contributors to these schemas and the researchers whose work underpins them.

The Cambridge Mathematics approach is based on the idea that exposure to mathematical concepts can pre-date formal learning, and this is particularly important in statistical literacy. The contextual nature of the subject allows students to use their prior knowledge and understanding of the 'story' of their data to undertake crude analysis and generate rudimentary diagrams that can be explored and refined into formal methods over time. Statistical literacy cannot be taught as a single lesson objective, but must be developed gradually over time through repeated exposure to, and refinement of, statistical methods as new data and scenarios are encountered that require more sophisticated skills.

At the heart of the Cambridge Mathematics Framework is the framework for design proposed by Burkhardt and Swann (see Table 1.2).

All the activities presented in Part 2 of this book will relate to one or more aspects of mathematics identified in the framework for design. By planning a series of activities that allow students to explore statistical techniques

Purpose of the lesson	Process genres on tasks – the student	Student products
Factual knowledge and procedural fluency	memorises and rehearses facts, procedures and notations	Performance
Conceptual understanding and logical reasoning	observes and describes phenomena	Description
	classifies and defines objects	Classification
	represents, and translates between representations	Representation
	justifies conjectures, procedures, connections	Explanation
	identifies and studies structure	Analysis
Problem solving and strategic competence	formulates models of situations	Model
	solves non-routine problems	Problem solution
	interprets and evaluates strategies	Critique

Table 1.2 Framework for design
[Adapted from *Design and Development for Large-Scale Improvement*, Burkhardt & Swann (2016)]

and which cover the full range of techniques identified by Burkhardt and Swann, students' developing statistical literacy can be supported and enhanced.

1.9 The place for formal techniques

Despite advocating an approach to statistical literacy that focuses on contextual analysis and critique, the mathematical skills of calculating statistics and representing data that are traditionally taught are still fundamental. To build a structure that will survive the test of time, both the bricks and the mortar that binds them together are needed – too little of either and the walls will collapse. Our aim in this book is to set out a balanced approach to learning about data to provide students with a solid understanding that will survive beyond formal education whether they are eventually expert, functional or occasional statisticians.

References and further sources

1. Kahneman, D. (2012). *Thinking, Fast and Slow* (London: Penguin).

2. Makar, K. & Rubin, A. (2014). Informal statistical inference revisited. *ICOTS-9*. <http://icots.info/9/proceedings/pdfs/ICOTS9_8C1_MAKAR.pdf> accessed 6th November 2017.

3. Steen, L. A. (Ed.) (1992). *Heeding the Call for Change: Suggestions for Curricular Action* (Washington, DC: Mathematical Association of America).

4. Garfield, J. (1995). How students learn statistics. *International Statistical Review/Revue Internationale de Statistique*, 63(1), 25–34.

5. Lightner, J. E. (1991). A Brief Look at the History of Probability and Statistics. *The Mathematics Teacher*, 84(8), 623–630.

6. Lehrer, R. (2007). Introducing students to data representation and statistics. *Mathematics: Essential for Learning, Essential for Life* (Proceedings of the 21st AAMT Biennial Conference), 22–41.

7. diSessa, A. A. (2004). Metarepresentation: Native Competence and Targets for Instruction. *Cognition and Instruction*, 22(3), 293–331.

8. Curcio, F. R. (1987). Comprehension of Mathematical Relationships Expressed in Graphs. *Journal for Research in Mathematics Education*, 18(5), 382–393. <https://doi.org/10.2307/749086> accessed 6th November 2017.

9. Franklin, C. A. (Ed.). (2007). *Guidelines for Assessment and Instruction in Statistics Education (GAISE) Report: A Pre-K–12 Curriculum Framework* (Alexandria, VA: American Statistical Association).

10 Burkhardt, H. & Swann, M. (2016). Design and Development for Large-Scale Improvement. *Proceedings of the 13th International Congress on Mathematical Education (ICME-13)*, 177–200.

You may find it interesting to look at some of the more holistic statistics curricula used in New Zealand and Australia:

http://nzcurriculum.tki.org.nz/The-New-Zealand-Curriculum/Mathematics-and-statistics/

http://victoriancurriculum.vcaa.vic.edu.au/mathematics

Chapter 2

The statistical cycle

2.1 Introduction

Due to its position in the mathematics curriculum, it is easy to see the essential skills of statistical analysis merely in terms of the mathematical processes that are essential when attempting to present data in a useable form. This perspective, though convenient for both teaching and assessment, fails to attend to many other elements that are essential to gain a working understanding of the subject. Statistical analysis does not simply involve throwing numbers at a set recipe of techniques, turning a handle and delivering an easy-to-understand outcome. If this were the case, technology is now such that human intervention would be just about unnecessary. Specialised computer statistics packages such as R and even simple spreadsheet software are now capable of generating summary statistics and graphical representations of data far more efficiently than humans can. (Chapter 8 introduces a classroom activity using R and Appendix 1 contains some useful R commands that can be used to enhance statistics lessons.)

Huge leaps forward are being made in machine learning to automate the analysis too, but, for the time being at least, human intervention is still essential both to select the appropriate form of the analysis and to interpret the output, even when using a computer. At the cutting edge of particle physics, the giant detectors in particle accelerators, such as the large hadron collider, generate so much data that the vast majority is discarded unanalysed; the decision of what to keep is made by computer algorithms that, although incredibly efficient, are not infallible.

> ### Large Hadron Collider
>
> In the first stage of the selection, the number of events is filtered from the 600 million or so per second picked up by detectors to 100 000 per second sent for digital reconstruction. In a second stage, more specialised algorithms further process the data, leaving only 100 or 200 events of interest per second. This raw data is recorded onto servers at the CERN Data Centre at a rate of around 1.5 CDs per second (approximately 1050 megabytes per second). Physicists belonging to worldwide collaborations work continuously to improve detector-calibration methods and refine processing algorithms to detect ever more interesting events.
>
> [from the CERN (2017) web page]

While computer algorithms are extremely effective at reducing the volume of data to more manageable levels, physicists must grapple with the challenge of filtering the data in a way that not only collects results that confirm existing theories but also captures unexpected results that may shed new light on their science. So many major steps forward in science have been the result of serendipity that relying on algorithms alone may risk some truly astounding signals getting lost in the noise.

Technology has rendered the 'by hand' production of summary statistics and graphical representations-trivial, but has in turn increased the importance of good decision-making both for the data fed in and for the resulting output. The mantra of computer programmers, 'rubbish in, rubbish out', has never been more appropriate. A computer will deal with whatever data is fed into it and produce statistical values even when the result is entirely inappropriate. Most of us have encountered this kind of issue when trying to draw a bar chart in a spreadsheet and generating a wholly incorrect diagram that uses what should be the criteria on the x-axis as a second data series (see Figure 2.1).

Number of siblings	0	1	2	3	4
Frequency	6	10	7	4	1

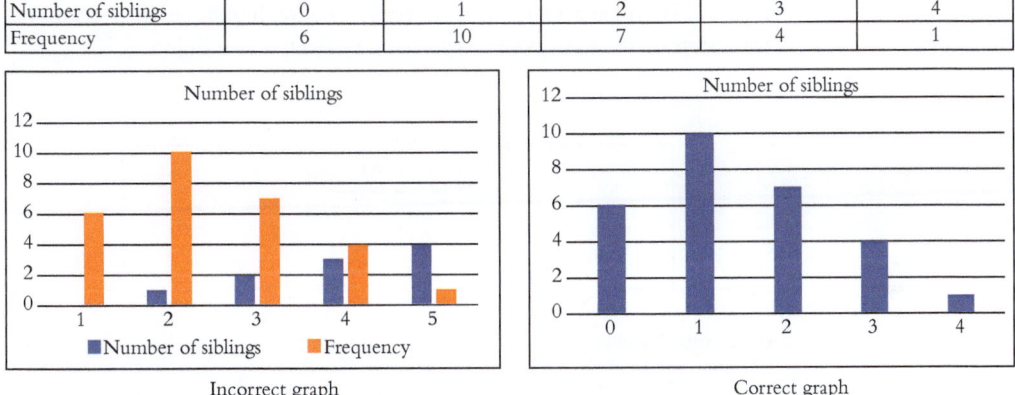

Figure 2.1 An example of the problems that can arise when using a spreadsheet to draw graphs

Increasingly, decision-making, analysis and interpretation are being pushed to the fore, making it essential that the more formal mathematical content of statistics is delivered in a way that maximises these aspects, while minimising the time spent on by-hand calculations and graph drawing.

The GAISE pre-K–12 framework (Franklin 2007) states that 'Statistical problem solving is a process that involves four components' and encourages the teaching of statistics within this framework. There is a growing consensus among education researchers that this approach should be adopted, with curricula beginning to make explicit reference to teaching statistical literacy as a complete cycle alongside the core mathematical techniques.

I. Formulate Questions
→ clarify the problem at hand
→ formulate one (or more) questions that can be answered with data

II. Collect Data
→ design a plan to collect appropriate data
→ employ the plan to collect the data

III. Analyze Data
→ select appropriate graphical and numerical methods
→ use these methods to analyze the data

IV. Interpret Results
→ interpret the analysis
→ relate the interpretation to the original question

Figure 2.2
[From *Guidelines for assessment and instruction in statistics education,* Franklin (2007)]

Presented as a list, these four steps can lose their cyclical nature (see Figure 2.2). In particular, the cycle version below highlights the return from step IV to step I which is so often neglected (see Figure 2.3).

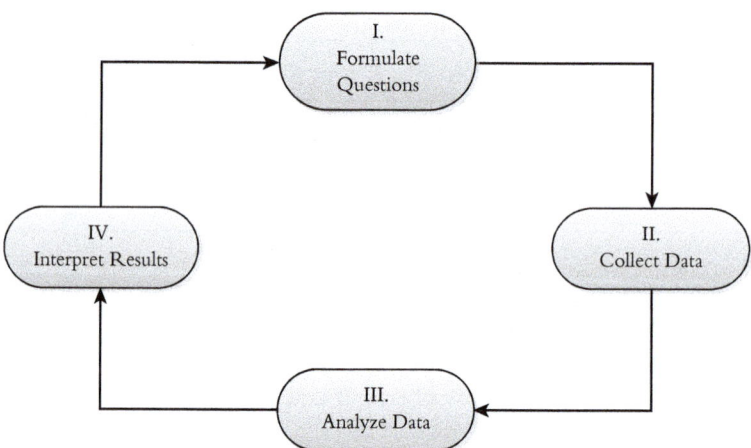

Figure 2.3 The statistical problem-solving cycle

Similarly, Teach-Stat, a CPD programme in the 1990s (Lajoie 2012), developed a concept map for the process of statistical investigation, setting out how the various techniques are tied together to make a coherent whole (see Figure 2.4).

At the centre, highlighted in red, are the same four steps as in GAISE (and most other representations of the cycle), but the concept map goes into much more detail about the processes within each step, their relationship to each other and their purpose within the statistical investigation.

These are just a few examples from a wealth of reports, teacher development programmes and academic papers that describe variations on a similar structure. However, despite the central position of this cycle in the development of statistical literacy and the inclusion of versions of the statistical cycle in many curriculum documents, the basic idea remains underrepresented in many classrooms.

2.2 Introducing the statistical cycle

When considering teaching a subject, it is helpful to ask the question 'Are we trying to produce drivers? Or passengers?' For someone who does not hold a driving licence it is entirely possible to spend many years as a passenger being transported to work, social events and local landmarks without ever knowing the locations of those places in relation to each other. A regular passenger may know how to get from home to work, or from home to the town centre, but have no concept of how to get directly between work and the town centre efficiently. A person who drives themselves, however, begins to understand their regular destinations not as discrete, unrelated places, but as locations on a larger map, and quickly learns how to get from one to another more directly (of course this analogy may have been more true before sat navs were widely available). Any course in statistics that seeks only to ensure that students are fluent in the techniques without attending to the landscape connecting them is neglecting arguably the most essential aspect of the subject.

It is helpful to begin to think about the practice of statistical analysis in the same context as science lessons. Few would suggest that the best way to teach students about experiments in science would be to separate all the individual skills involved in a practical experiment and teach them separately, without connecting them together as a complete process.

Statistical literacy should be approached in the same way, beginning with a question to be answered and working from there to a conclusion, or even a new question. This is not to suggest that every single lesson should involve students working through the complete cycle, or even multiple cycles, but it means that calculation without context should be avoided. Instead, wherever possible, teachers should seek opportunities to develop new skills within the framework of the statistical cycle.

As students progress through school, they will develop more sophisticated skills which they can apply to statistical questions. However, with support, the complete statistical cycle is something that can be applied at all stages of schooling. An early experience of statistics aimed at primary school students could begin with the question 'What type of pet is most popular in our class?' Teachers can model the cycle on behalf of the students, eliminating the need for students to have all the necessary skills before they begin working with a statistical problem.

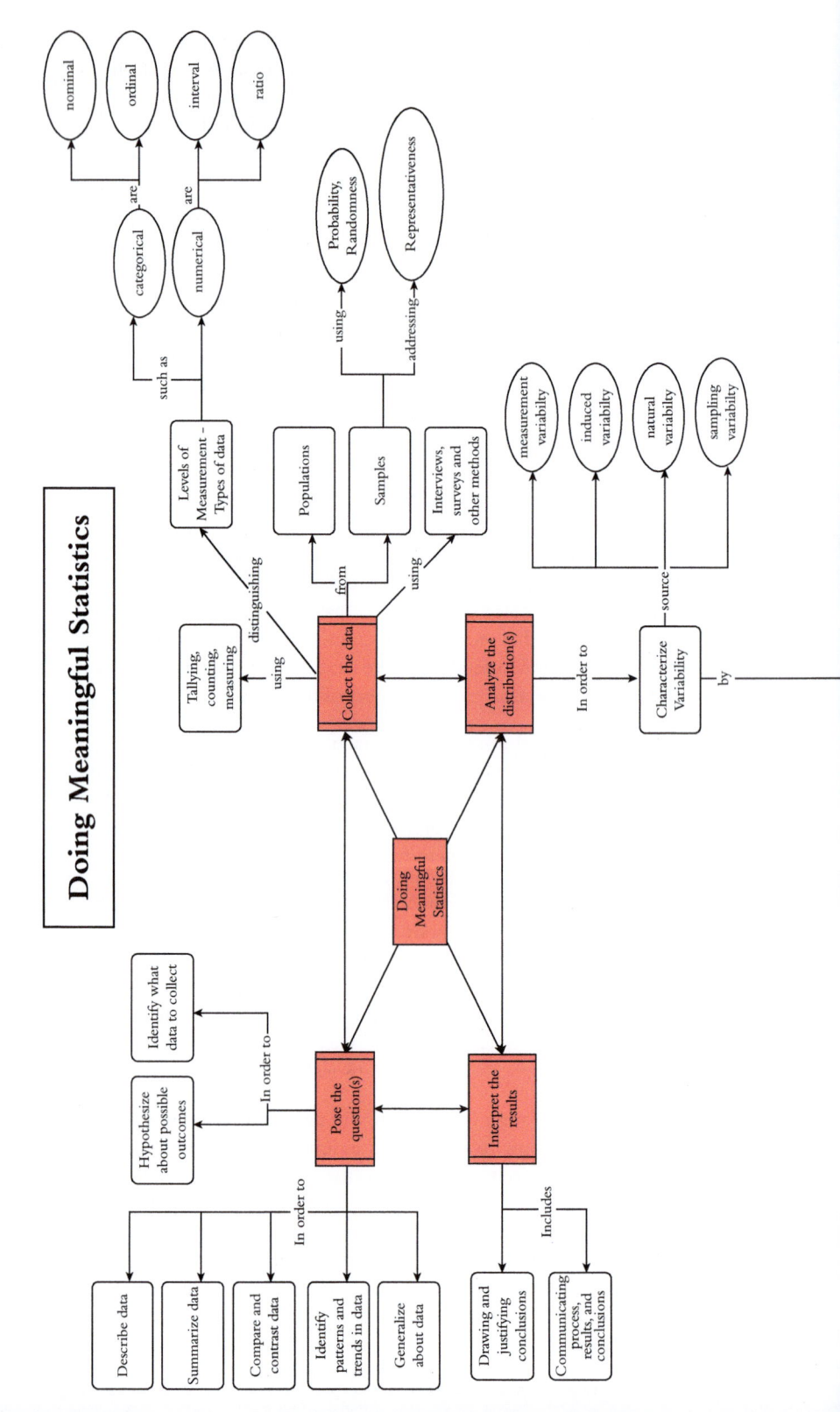

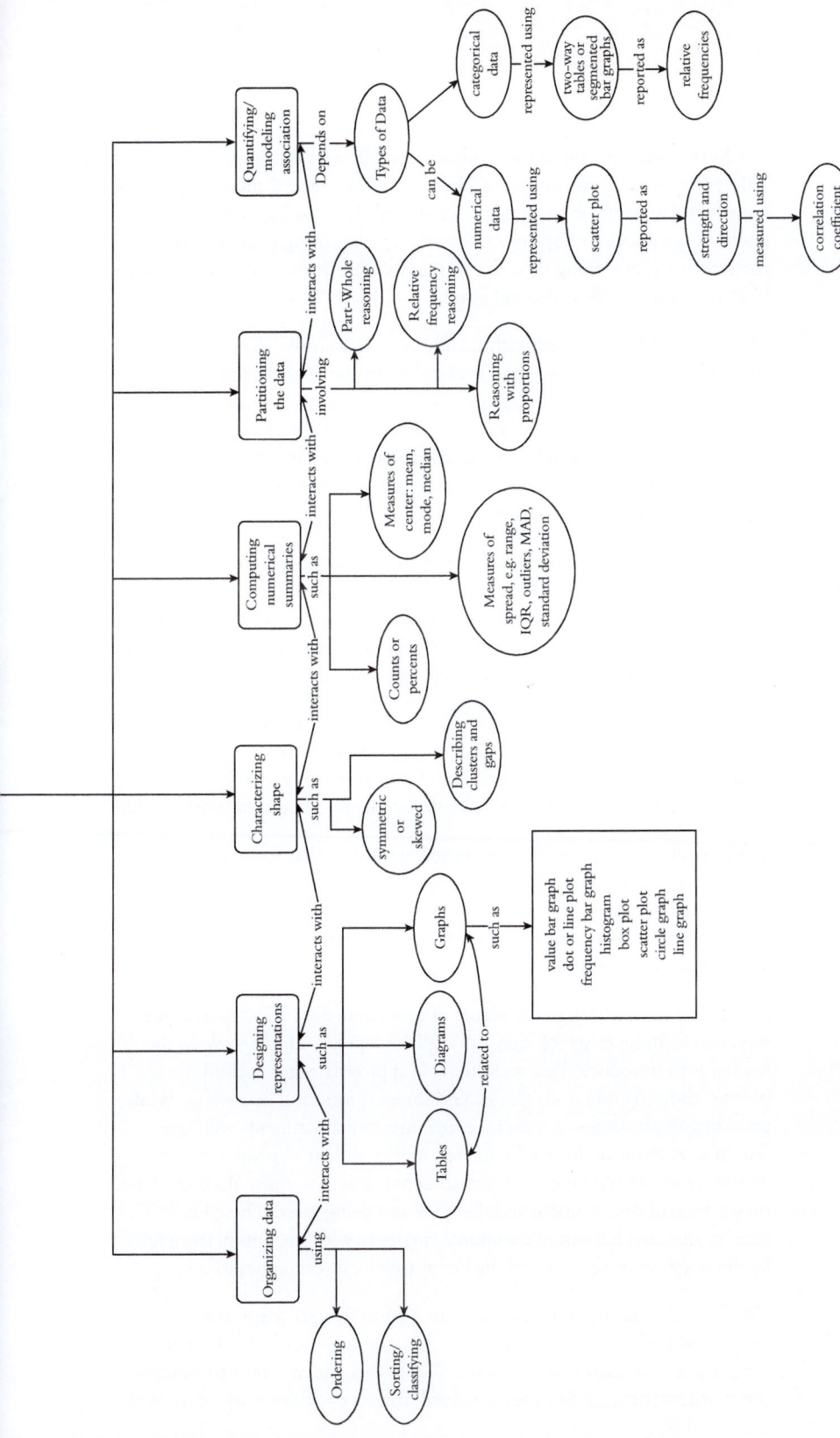

Figure 2.4 Concept map for statistical investigations

Firstly, the question can be set and students allowed to shout out the different types of pets they have. The teacher can record these, and ask for a show of hands for each category of animal. The results, once recorded, could be graphed, perhaps electronically or even using more creative methods such as getting students to stand with others who have the same kind of pet to create a physical graph.

Issues such as how to deal with people who have more than one pet should be discussed rather than avoided, with the class deciding, with justification, the rules to be applied. Should all the pets be included? Should each student choose their favourite pet? What about students who have no pets – could they choose the pet they would most like to have? Is a bar graph even the most appropriate representation in this instance? What about a Venn diagram …

If appropriate for the class, smaller groups of students could be allowed to decide on their own set of rules before each group collects data from the same population (their class). This will introduce an element of variation in the data that students have collected, encouraging them to think about this fundamental element in a relatable context.

Once a decision has been taken, with justification, as to the most popular pet, teachers should encourage students to critique the whole process: would they make decisions differently if they were to repeat the process? What could they have done to make data collection easier or more reliable?

While rudimentary, this kind of approach ensures that students' initial experiences of statistics encompass the full cycle. It also allows students to begin to recognise some of the issues that could arise during data collection and interpretation, and to develop strategies for anticipating and dealing with them.

As students move through the phases of education, they will develop more skills and techniques which they can apply independently. The role of the teacher is to introduce these techniques as appropriate and scaffold the process, choosing questions that extend students' statistical knowledge while providing opportunities for decision-making. Initially students will have few choices to make due to the limited toolkit at their disposal, but as they progress, they should be provided with opportunities to make decisions based on the type of data available and the questions being asked. These kinds of choices are often left out of traditional curricula, with students being guided by the question or the topic of the day as to which techniques to use.

Consider the question reproduced in Figure 2.5; students are instructed precisely how to use the data and are clearly told what form their conclusion should take. While this allows them to practise the mathematical techniques involved, it fails to address any statistical process skills.

8 Amber measures the heights of some young trees and the widths of their trunks. The results are shown in the table below.

Width of trunk (cm)	10	11	12	14	18	19	22	23	28	29
Height of tree (m)	4.5	5.5	7.5	12	3.5	12.5	11.5	16	15	18

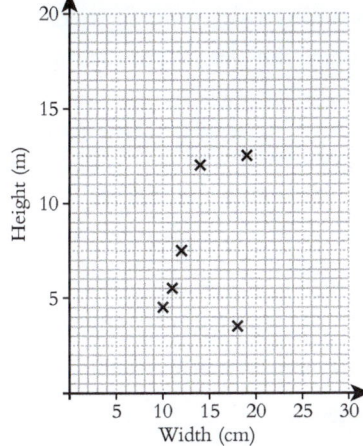

(a) The first six points have been plotted on the scatter diagram.
Complete the diagram by plotting the last four points. **[2]**

(b) State the correlation shown by the scatter diagram.

(b) _____ **[1]**

(c) Use your diagram to describe the relationship between the width of a tree trunk and the height of the tree.

_____ **[1]**

(d) (i) Draw a line of best fit on the diagram.

(ii) Amber has a tree with a trunk width of 25 cm.
Use your diagram to estimate the height of this tree.

(d)(ii) _____ **[1]**

(e) One of these trees is from a different species.
On the diagram put a circle around the point for that tree. **[1]**

Figure 2.5 Traditional exam question
[OCR, GCSE Mathematics, Paper J567/03, June 2015]

The same question can be amended to address the full statistical cycle by providing the data and setting something more interpretative and discursive, for example:

> Amber wants to know if there is any relationship between the age of young trees and the width of their trunks so that she can predict the age of a tree in her garden with a trunk of width 25 cm.
>
> **a** Use the data in the table to provide Amber with an answer and explain your reasoning.
>
> **b** Do you think all trees in the data were of the same species? Justify your answer.

The same mathematical skills are practised, but students must attend to other key aspects too, such as

- interpreting how the data can be used to answer the question
- selecting the appropriate analysis technique
- drawing the graph
- identifying how to interpret the graph, in this case drawing a line of best fit if appropriate
- commenting on the relationship and explaining how it relates to the original question
- identifying any outliers and considering the implications.

There is also room for the more proficient students to engage in analysis that goes beyond the requirements of the original question. While this might not be feasible in assessment terms, within the classroom questions which are a little more open-ended (or open-middled) can provide good opportunities for growth. You could, for example, ask about ways in which Amber could improve her data collection or focus her question more carefully.

When teaching statistics, consideration must be given to the wider cycle, ensuring that the skills students are practising are embedded within a context which enables decisions to be made. Students unused to working in an environment where they are encouraged to formulate questions, make decisions, select techniques and draw conclusions are likely to require significant support, especially early on. As a result, it will almost certainly be necessary for teachers to scaffold students' early experience of the cycle. This can be easily achieved by choosing a specific stage of the cycle to focus on and structuring lessons so that students are given time and freedom to work on this aspect. Other stages of the cycle can then be entirely modelled by the teacher, or supported by leading questions within activities. On occasions where there is more classroom time available, and as students become more experienced, teachers can allow students more space and freedom to work through additional stages unaided.

2.3 Formulate questions

One of the most neglected parts of the statistical cycle, formulating questions, is the precursor to experimental design and an essential part of the process of understanding the need for data. Students need to understand the difference between mathematical questions with a deterministic answer and statistical questions that require more data collection and interpretation of summary values or distributions.

'How old am I?' is a straightforward question to answer and leads to a single value. A statistical question 'Is my height typical of the heights of people in my class?' requires more data and much deeper consideration.

> **How tall am I?**
>
> 'How tall am I?' is an interesting question as it can be considered deterministic if the method of measurement is highly accurate; however, it can lead to some wild variation in values if the method of measurement is less precise. Asking individual students to all measure the height of the teacher in turn using inefficient methods of measurement such as a short ruler can lead to an interesting data set and allow classes to begin exploring how to define signals within the noise. Is it possible to identify the teacher's true height in the distribution of inexactly measured values?

The purpose of encouraging students to identify questions to ask is to help them gain an understanding of the difference between mathematical questions and statistical questions as well as an appreciation of how data can be used to explain and organise the world around them. The GAISE report sets out stages for development of statistical literacy, suggesting that questions be initially limited to things that can be explored from data collected within the classroom, before moving on to data collected within the wider environment of the student and beyond.

With this in mind, asking students to investigate aspects of themselves may prompt many questions that can be explored by statistical methods, for example:

- Am I typical of students in my class?
- What pets are most popular in my class?
- How far do students in my class live from school?
- How good are the reactions of students in my class?

You may have noticed that every question above contains the words 'in my class'. Even before they begin to consider the techniques of sampling and inference, students can develop a sense of the limit of their statistical conclusions. Encouraging students to define the population within their question is an effective way of imposing a limit on the conclusions they draw; for example, the results of the investigations above will be valid only when considering the class in question. Later it will be easy to extend this idea by asking students to consider what the class data allows them to infer about their year group, their school, their town or the country. Students can then begin to explore sampling from a larger population and whether their conclusions can be considered representative.

To guide the creation of questions, teachers may provide some sort of stimulus to lead students in a certain direction, for example providing tape measures and asking students to come up with a list of questions they could investigate using only that piece of equipment, or providing an article about an interesting phenomenon or theory like Vitruvian man that can be explored in class and asking how students would test the accuracy of the claims made. Another option might be to begin with online data visualisations such as the Constituency Explorer (www.constituencyexplorer.org.uk/), which provides data from the UK Office for National Statistics collected during recent general elections.

Exploring real data provides an opportunity to overcome various cognitive biases known as 'heuristics'. The availability heuristic tells us that people tend to overestimate the likelihood of things that are easy to recall, and the representativeness heuristic tells us that people attend to characteristics that fit their mental templates, without taking into account the prevalence of such characteristics in the wider population. Data from sources such as Constituency Explorer allows students to question their own perceptions by exploring the data presented in an easy-to-digest form, and can be used to prompt further questions related to their own class.

2.4 Collect data

Once students have formulated a question to answer, they can begin to consider the mechanics of collecting the data necessary to find an answer. This can be a messy and time-consuming process, and careful choice of questions (with guidance from a teacher) is necessary to ensure that this part of the cycle runs smoothly. However, the benefits outweigh the logistical drawbacks, and a well-chosen task can boost student engagement, with the associated benefits to learning.

For early experiences, single-question surveys of members of the class are sufficient to produce useable frequency data, and for time-series data, longer running experiments could be set in motion in advance of the key lesson. Students could be asked to record the weather every day for several weeks, or they could plant seeds in a pot at the beginning of term and measure the heights of the growing plants on a daily or weekly basis. Another option is to spend a lesson collecting and recording multivariate student data early in the school year, including a full range of data types. Teachers and students can then return to this data regularly, selecting the appropriate subset of data to answer a specific question designed by the teacher in order to practise a specific technique. Another approach would be to ask a specific question, provide students with a secondary data source and set them the task of identifying and extracting the necessary information to answer the question. (Multivariate data sets are covered in more detail in Chapter 3.)

2.5 Analyse data

Analysis of data is probably the aspect of the statistical cycle that most teachers are most comfortable with, as traditional textbooks focus predominantly on graphical representations and calculations with data. Within the statistical cycle, however, students must make decisions for themselves about which techniques to employ, potentially trying several and discarding those that prove ineffective. An excellent example of the type of activity that generates this kind of decision-making is detailed in *How faithful is Old Faithful* (Pfannkuch & Shaughnessy 2002), concerning the time between eruptions of the geyser 'Old Faithful'. Approaching the data set in a formulaic way by calculating an average for all the data and constructing a histogram, cumulative frequency chart or boxplot will provide students with some interesting things to interpret. However, if the time between eruptions is also plotted as an ordered series of points on a line graph with time between eruptions on the vertical axis, a new feature of the data becomes visible, namely that there is a cyclic pattern of long–short intervals taking place. This would have been difficult to spot if only a single type of representation was used.

> **A note on meta-representational competence (MRC)**
>
> Meta-representational competence (diSessa 2004) encapsulates the ability of students to invent representations of data on the fly, critique the effectiveness of representations, articulate the purpose of representations in context of the data, and learn new types of representation rapidly. Advocates of MRC encourage teachers to allow students to create and refine meta-representations of data rather than teaching sanctioned graphical forms, with these ongoing refinements eventually leading to establishing widely understood

> conventions, such as 'increasing value' being best represented by 'up' in relation to the y-axis. There is some suggestion in research that the ability to create representations is linked to general problem-solving ability and more efficient assimilation of new graphing techniques. Ideally this exploration of representations begins in early schooling and continues throughout a student's education across a range of topics in mathematics, and in other subjects where appropriate.

2.6 Interpret results

The quality of interpretative work is heavily dependent on the quality of the analysis, and as students gain new tools in their toolkit, increasingly sophisticated responses will be available to them. For interpreting graphs, various frameworks exist to describe this sophistication (Friel, Curcio & Bright 2001), and while there are some differences in how the processes are described and categorised, they generally lie on a similar trajectory.

> Stage 1: Extracting information from the data.
> Stage 2: Finding relationships in the data.
> Stage 3: Moving beyond the data.

Traditional exercises in statistics often focus on the most elementary aspects of interpretation, asking students to read values from graphs (stage 1) or to describe trends and patterns in terms of the graph itself by identifying associations, or asking comparative 'how many more?' questions (stage 2). Attention must also be given to stage 3, moving beyond the data and relating the visible features of the graph to the context in which the data exists. Students find this incredibly challenging, but if initial statistical investigations are confirmatory (asking students to use data to demonstrate things they already know to be true), then students' prior deep understanding of the context can be explored. This will enable them to relate patterns and significant information they are already aware of to features of graphs and statistical measures.

As students progress through secondary education, the role of probability in statistics becomes increasingly important. Using statistical enquiry to help students develop a firm foundation in the non-deterministic nature of probability provides a straightforward way for students to engage with the statistical cycle, particularly when probability is taught in terms of expected frequencies. Useful experimental data can be quickly generated, recorded, graphed and discussed in a straightforward context. *Teaching Probability* by Jenny Gage and David Spiegelhalter, also in this series, sets out a detailed approach to teaching probability in this way and provides several experimental activities that can be conducted in any classroom.

2.7 Measuring progress

The NCTM standards for statistics and probability (National Council of Teachers of Mathematics 2017) provide a detailed framework for statistical enquiry, with statistical techniques common to many international curricula presented as steps in the statistical cycle (see Table 2.1).

NCTM Data Analysis and Probability Standards
1. Formulate questions that can be addressed with data and collect, organize, and display relevant data to answer them

Pre-K–2 Expectations: In pre-kindergarten through grade 2 all students should–	Grades 3–5 Expectations: In grades 3–5 all students should–	Grades 6–8 Expectations: In grades 6–8 all students should–	Grades 9–12 Expectations: In grades 9–12 all students should–
• pose questions and gather data about themselves and their surroundings; • sort and classify objects according to their attributes and organize data about the objects; • represent data using concrete objects, pictures, and graphs.	• design investigations to address a question and consider how data-collection methods affect the nature of the data set; • collect data using observations, surveys, and experiments; • represent data using tables and graphs such as line plots, bar graphs, and line graphs; • recognize the differences in representing categorical and numerical data.	• formulate questions, design studies, and collect data about a characteristic shared by two populations or different characteristics within one population; • select, create, and use appropriate graphical representations of data, including histograms, box plots, and scatterplots.	• understand the differences among various kinds of studies and which types of inferences can legitimately be drawn from each; • know the characteristics of well-designed studies, including the role of randomization in surveys and experiments; • understand the meaning of measurement data and categorical data, of univariate and bivariate data, and of the term variable; • understand histograms, parallel box plots, and scatterplots and use them to display data; • compute basic statistics and understand the distinction between a statistic and a parameter.

Table 2.1 NCTM data analysis and probability standards
[Reprinted with permission from Principles and Standards for School Mathematics, copyright 2000 by the National Council of Teachers of Mathematics (NCTM).]

NCTM Data Analysis and Probability Standards			
2. Select and use appropriate statistical methods to analyze data			
• describe parts of the data and the set of data as a whole to determine what the data show.	• describe the shape and important features of a set of data and compare related data sets, with an emphasis on how the data are distributed; • use measures of center, focusing on the median, and understand what each does and does not indicate about the data set; • compare different representations of the same data and evaluate how well each representation shows important aspects of the data.	• find, use, and interpret measures of center and spread, including mean and interquartile range; • discuss and understand the correspondence between data sets and their graphical representations, especially histograms, stem-and-leaf plots, box plots, and scatterplots.	• for univariate measurement data, be able to display the distribution, describe its shape, and select and calculate summary statistics; • for bivariate measurement data, be able to display a scatterplot, describe its shape, and determine regression coefficients, regression equations, and correlation coefficients using technological tools; • display and discuss bivariate data where at least one variable is categorical; • recognize how linear transformations of univariate data affect shape, center, and spread; • identify trends in bivariate data and find functions that model the data or transform the data so that they can be modelled.

Table 2.1 (*continued*)

NCTM Data Analysis and Probability Standards			
3. Develop and evaluate inferences and predictions that are based on data			
• discuss events related to students' experiences as likely or unlikely.	• propose and justify conclusions and predictions that are based on data and design studies to further investigate the conclusions or predictions.	• use observations about differences between two or more samples to make conjectures about the populations from which the samples were taken; • make conjectures about possible relationships between two characteristics of a sample on the basis of scatterplots of the data and approximate lines of fit; • use conjectures to formulate new questions and plan new studies to answer them.	• use simulations to explore the variability of sample statistics from a known population and to construct sampling distributions; • understand how sample statistics reflect the values of population parameters and use sampling distributions as the basis for informal inference; • evaluate published reports that are based on data by examining the design of the study, the appropriateness of the data analysis, and the validity of conclusions; • understand how basic statistical techniques are used to monitor process characteristics in the workplace.

Table 2.1 (*continued*)

While aiming to describe the essential components of a statistics programme, these NCTM standards can be used as a framework to assess students' ability in statistics. As an alternative to traditional exam-based assessment, a stimulus or question could be given to students to investigate using data either provided or collected depending on available time and resources.

Students' work can then be assessed against each of the four aspects of the statistical cycle. The statements prescribed by the NCTM at different stages of the curriculum can be used as competence indicators, with some small adjustments to take into account jurisdictional context and curricular restrictions.

Historically in the UK, statistical investigations were assessed as part of the GCSE through an independent coursework task. Students were provided with a stimulus to investigate and their work was marked against a published assessment framework. While no longer part of the UK GCSE, these tasks still provide excellent stimulus material for assessment of students' statistical abilities. A selection of such tasks can be found in Appendix 2.

The importance of context in statistics should never be underestimated. While it is possible to provide some contextual information in a short question, it is difficult for students to engage fully with the information and consider the deeper implications within any scenario provided. This is more commonly recognised in non-mathematical subjects, where stimulus material is routinely provided in advance of assessment so that students can familiarise themselves with it before answering related questions in an exam. Working through the whole statistical cycle provides a deeper understanding of the context which allows for more sophisticated interpretation and inference following any analysis. It also gives students a clearer understanding of the circumstances under which various statistical techniques should be employed.

Familiarity with the kinds of interpretations and inferences that can be made when working in a secure context can then provide a template for problem solving within the less familiar and less well-defined contexts found in the wider world beyond education.

References and further sources

1. CERN (2017). Processing: What to record? <https://home.cern/about/computing/processing-what-record> accessed 6th November 2017.

2. Franklin, C. A. (Ed.). (2007). *Guidelines for Assessment and Instruction in Statistics Education (GAISE) Report: A Pre-K–12 Curriculum Framework* (Alexandria, VA: American Statistical Association).

3. Lajoie, S. P. (2012). *Reflections on Statistics: Learning, Teaching, and Assessment in Grades K–12* (Routledge).

4. PfannKuch, M. & Shaughnessy, J. M. (2002). How Faithful Is Old Faithful? Statistical Thinking: A Story of Variation and Prediction. *Mathematics Teacher*, 95(4), 252–259 (National Council of Teachers of Mathematics). <http://www.nctm.org/Publications/mathematics-teacher/2002/Vol95/Issue4/How-Faithful-Is-Old-Faithful_-Statistical-Thinking_-A-Story-of-Variation-and-Prediction> accessed 6th November 2017.

5. diSessa, A. A. (2004). Metarepresentation: Native Competence and Targets for Instruction. *Cognition and Instruction*, 22(3), 293–331.

6. Friel, S. N., Curcio, F. R. & Bright, G. W. (2001). Making Sense of Graphs: Critical Factors Influencing Comprehension and Instructional Implications. *Journal for Research in Mathematics Education*, 32(2), 124–158. <https://doi.org/10.2307/749671> accessed 6th November 2017.

7. Gage, J. & Spiegelhalter, D. (2016). *Teaching Probability* (Cambridge University Press).

8. National Council of Teachers of Mathematics (2017). Standards and Positions: Principles and Standards for Data Analysis and Probability. <http://www.nctm.org/Standards-and-Positions/Principles-and-Standards/Data-Analysis-and-Probability> accessed 6th November 2017.

Chapter 3

Exploratory data analysis

'Data analysis must be willing to err moderately often in order that inadequate evidence shall more often suggest the right answer.'

Tukey 1961

3.1 Introduction

As discussed previously, we live in a world with very large amounts of data and with sophisticated technology and software with which to interrogate that data. To some extent, then, the nitty-gritty of applying techniques to the data is not an issue; rather it is the decisions about how to undertake the analysis and, most importantly, what questions to ask that are critical. This hints at another shift in the way that statistical thinking has developed – away from asking a question and collecting relevant data for that question, towards collecting large (sometimes large enough to be described as 'big') amounts of data and looking at these large data sets to suggest interesting questions. This is where exploratory data analysis comes in.

3.2 What is exploratory data analysis?

Exploratory data analysis is the name given to a collection of ways of working which aim to get away from the constraints of formal modelling or hypothesis testing. It refers to the use of data summary and presentation techniques in order to draw informal inference and perhaps motivate further data collection or the application of hypothesis tests. Some of the defining characteristics of the approach are the use of measures of location and spread that control for skew and outliers, such as the median and quartiles, the plotting of small multiples (see Section 3.5) to make sense of complex multivariate data sets, and more recently the use of dynamic and interactive representations of data.

The use of exploratory data analysis was first popularised by John Tukey in *The future of data analysis* (Tukey 1961), and was also described in great detail in the wonderful *Exploratory Data Analysis* (Tukey 1977). Tukey extolled the virtues of flexibility of approach, of finding ways for data to express their apparent character and of using graphs as tools for indicating phenomena.

Tukey also looked forward to the opportunities afforded by computing – to a time when graphs would be drawn by computers, when experimental sampling would become practical, and even when the machine would do most of the work and a human would simply be needed to supervise and approve its conclusions. It is certainly the case that in the past, much dynamic exploration of statistics was simply impossible because we were confined to processes and activities that were computable by hand. We are much less constrained by the computable now, and this opens up exploratory data analysis to a much wider group of students; however, in the classroom there are often barriers to do with access to, and familiarity with, technology. Nevertheless, it is worth remembering the advice of David Moore (1992) that 'the teaching of statistics should … mirror statistical practice, where computation and graphics are almost entirely automated', and of George Cobb (2007): 'Technology allows us to do more with less: more ideas, less technique. We need to recognise that the computer revolution in statistics education is far from over.'

The traditional or classical statistical approach is called 'confirmatory' by Tukey to highlight that it is centred around forming hypotheses and then confirming them (or not) through the application of a toolbox of techniques for specific situations. Note also that the confirmatory approach involves (sometimes implicitly) setting up models for the data, based on some suitable distribution along with some parameters which we either know or estimate, and then using those models. The layers of abstraction in this process are many, making access difficult for students beginning their studies in statistical literacy. In exploratory data analysis, no such assumptions are made; we can jump straight into the data to see what is there, rather than attempting to model it.

Finding good questions is just as important as finding good answers

Exploratory data analysis (let's look at the data and see if it suggests any interesting relationships or patterns) does not take the place of confirmatory data analysis (I think this is true, so let's find out with a hypothesis test). Rather they complement each other and together form a strong foundation for analysis.

As an investigatory tool, exploratory data analysis allows for the discovery of patterns which were not part of the experimental design and which might suggest further avenues for research and data collection. For example, imagine some researchers who are looking into what variables affect the percentage of people travelling to work by bicycle. They might think that large populations lead to better provision of infrastructure for cyclists and indicate higher levels of urbanisation. This thought process might lead them to looking at something like the UK census data and plotting a scatter diagram (see Figure 3.1). However, doing so immediately throws up some interesting anomalies, and even an apparently simple diagram like the one in Figure 3.1 can be a rich source of discussion in the classroom.

Without carrying out any formal analysis at all, this could lead to discussions about why or how some local authorities have very high

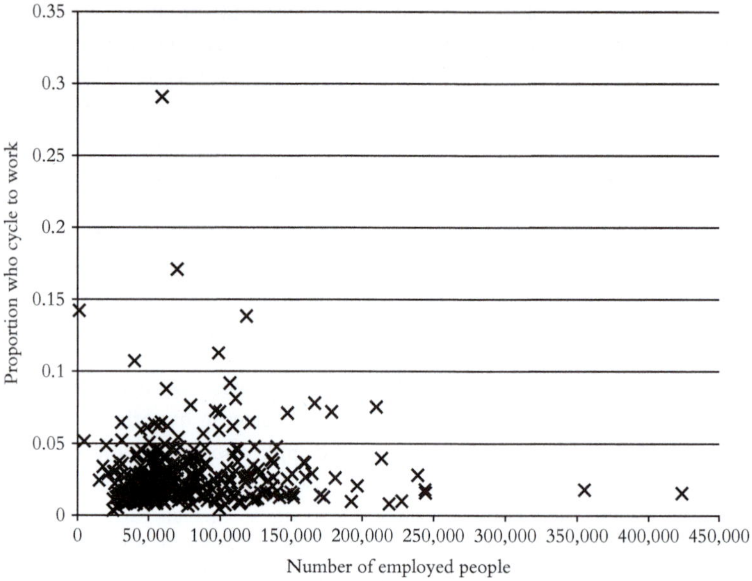

Figure 3.1 Scatterplot showing the number of people in employment in each local authority in England and Wales and the proportion of those people who cycle to work, according to the 2011 UK Census [Adapted from data from the Office for National Statistics, www.ons.gov.uk]

rates of cycling to work, to the realisation that not all very large local authorities are cities, or to why there is such a large 'blob' of data in the bottom left. Intuitive thinking about the patterns might lead to the idea that while there is no direct relationship between the variables, almost all local authorities have fewer than 1 in 20 people cycling to work; but there seems to be a subpopulation of small local authorities that behave in a different way to all of the others, with the highest proportion of people cycling to work in some of the smallest local authorities.

A little analysis, such as five-number summaries and boxplots along the axes, will highlight just how many data points there are in the blob (there are 348 local authorities in the data set), which tells us that the outliers are even more unusual than the diagram initially suggests. Attempting to add a line of best fit is not likely to get us very far, but is there a curve that might fit the data? Does the fact that you can fit the reciprocal function well mean that there is a useful model based on this? What does this mean for our lines of best fit in other contexts and the strength of the conclusions we draw from them?

The five-number summary

The five-number summary is the name given to a useful collection of five easily calculated statistics that can be used to describe or define a distribution. It consists of the smallest value, the lower quartile, the median, the upper quartile, and the largest value. These numbers will be familiar to teachers as they make up the five key points of a boxplot.

If nothing else, a scatter diagram such as this is a wonderful opportunity to discuss outliers as an indication of variability, not just errors. These outliers are not incorrect values, so while dismissing them to look at the 'normal' areas might allow us to infer something about general trends, it misses the opportunity to look at the really interesting areas. Why are they so far away from the norm?

All of this is exploratory data analysis. It is easy to see how just presenting (or creating) a single rich diagram can generate questions and allow for informal analysis, formal analysis, discussion of statistical concepts, further non-mathematical investigations, further statistical analysis, discussion of data presentation and so on.

3.3 Teaching exploratory data analysis

So what techniques do students need to learn to allow them to do this? John Tukey, populariser of EDA (exploratory data analysis) in the 20th century, was clear: 'Exploratory data analysis is an attitude, a flexibility, and a reliance on display, NOT a bundle of techniques, and should be so taught' (Tukey 1980). In terms of pedagogy, the essential idea (Tukey 1961) is to teach students 'not "what to do", nor "how we learned what to do", but rather "what we have learned".' This requires a fundamental change in the attitude towards data analysis. Instead of starting with trying to nail down what the data establish, we should begin with free use of ad hoc and informal techniques to simply see what the data suggest. It also requires us as teachers to be brave about context – using realistic problems, asking potentially difficult questions and being open about the approximate and inferential nature of conclusions.

In the context of the cycling investigation in the previous section, there is an opportunity to practise techniques such as finding correlation coefficients and five-number summaries, and to consider outliers. It also provides scope for discussing ideas of association, correlation, subpopulations and robustness. However, the focus should be on *what has been learnt*, not only about the data set, but also about careful use of the techniques. There are of course a very large number of highly sophisticated tools that we could use for exploratory data analysis, but Tukey is clear that our simple tools are up to the task.

Before going any further, it is important to note two pitfalls when considering exploratory data analysis. One of the biggest problems in interpreting data, particularly informally, is not ignorance but preconceived ideas; it is very easy to see in the data the pattern you want, or expect, to see. In the classroom you can avoid this to some extent by making sure that students work sometimes independently and sometimes in groups, that they get used to presenting and justifying their conclusions to each other, and most importantly that they feel comfortable with challenging each other's reasoning.

The second pitfall is a bit more technical, but no less important. Using both confirmatory and exploratory data analysis on the same set of data is potentially very dangerous, leading to systematic bias. Discussing why this is the case would be a useful classroom activity in itself, particularly following an exercise in which both approaches were taken. The problem with using the data that suggests a given relationship to confirm the existence of that relationship is not necessarily obvious.

3.4 Robust measures

Classical statistical measures such as the mean and standard deviation are often highly sensitive to outliers, measurement errors, missing data and so on. Robust measures, such as the median, seek to get around these problems by using procedures which minimise the effect of errors and outliers. These measures are therefore not susceptible to sampling variation and do not require careful cleaning of data. This last aspect is becoming more and more important as data sets get bigger and bigger, and where we are often using secondary data. In the modern world of big data, the usefulness of exploratory data analysis as a quick way of looking for patterns and questions is ruined if you have to go through careful cleaning and analysis before starting. However, some caution is needed, as the cycling example shows; what sort of conclusions might we have reached by simply finding the mean and standard deviation of each measure, or even the equation of a regression line?

The five-number summary is a standard set of simple statistics used to describe the *distribution* of a data set. The italics are important; the five-number summary includes information about the location and spread of the data, and it does so using robust measures.

The five numbers are the sample minimum, the lower quartile, the median, the upper quartile and the sample maximum. Obviously the first and last are very sensitive to outliers, but the middle three are robust. Because they are order statistics, changing the extreme values, or even corrupting values in the middle of the range, has little or no effect on these measures, as shown in the activity at the end of this section.

So, just how resistant is the median? Because it is an order statistic, errors in the collection and recording of data for up to 50% of the dataset can still result in the median value being in the right ballpark. You can make any values above (or below) the median arbitrarily large (or small) without it moving at all! Other kinds of strangeness in distributions which lead to unrepresentative values of the mean, such as fat tails, high skew or long tails, are also smoothed by the median. In fact, you do not even need to know what the extreme values in a data set are in order to be able to find an accurate value of the median, so in cases where the extreme values show a high degree of sampling variation, or where they can be difficult to measure accurately, the median can still be used as a good measure of location.

The quartiles also have these properties, though of course to a lesser degree. You can also use them to find the interquartile range, which is a robust measure of spread. As with the mean, the standard deviation of a data set can be made arbitrarily large (or small) by changing a single value in the data set, but this is not a problem for the interquartile range. Even if you corrupt a value between the quartiles, it is only if the changed value goes outside them that the interquartile range might change, and it makes no difference how large or small that value becomes.

A great activity to demonstrate the utility of the five-number summary is based on the classic 'Anscombe's quartet' (see Tables 3.1 and 3.2). These four bivariate data sets are highly synthetic, but they provide an excellent demonstration of the importance of graphing and the effect of outliers on classical measures. Frank Anscombe created these data sets (Anscombe 1973) to attack the idea that the calculations of summary data are somehow 'exact' but the graphs are somehow 'inexact', and also to demonstrate the effect of outliers on summary measures. The four sets have identical mean and variance (and a host of other classical summary measures including correlation coefficients and the equations of the

1		2		3		4	
x	y	x	y	x	y	x	y
10.0	8.04	8.0	6.58	10.0	7.46	10.0	9.14
8.0	6.95	8.0	5.76	8.0	6.77	8.0	8.14
13.0	7.58	8.0	7.71	13.0	12.74	13.0	8.74
9.0	8.81	8.0	8.84	9.0	7.11	9.0	8.77
11.0	8.33	8.0	8.47	11.0	7.81	11.0	9.26
14.0	9.96	8.0	7.04	14.0	8.84	14.0	8.10
6.0	7.24	8.0	5.25	6.0	6.08	6.0	6.13
4.0	4.26	19.0	12.5	4.0	5.39	4.0	3.10
12.0	10.84	8.0	5.56	12.0	8.15	12.0	9.13
7.0	4.82	8.0	7.91	7.0	6.42	7.0	7.26
5.0	5.68	8.0	6.89	5.0	5.73	5.0	4.74

Tables 3.1 Anscombe's quartet of data sets

	1		2		3		4	
	x	y	x	y	x	y	x	y
Mean	9	7.50	9	7.50	9	7.50	9	7.50
Sample variance	11	4.125	11	4.125	11	4.125	11	4.125

Tables 3.2 Mean and variance of the four sets of data in Anscombe's quartet

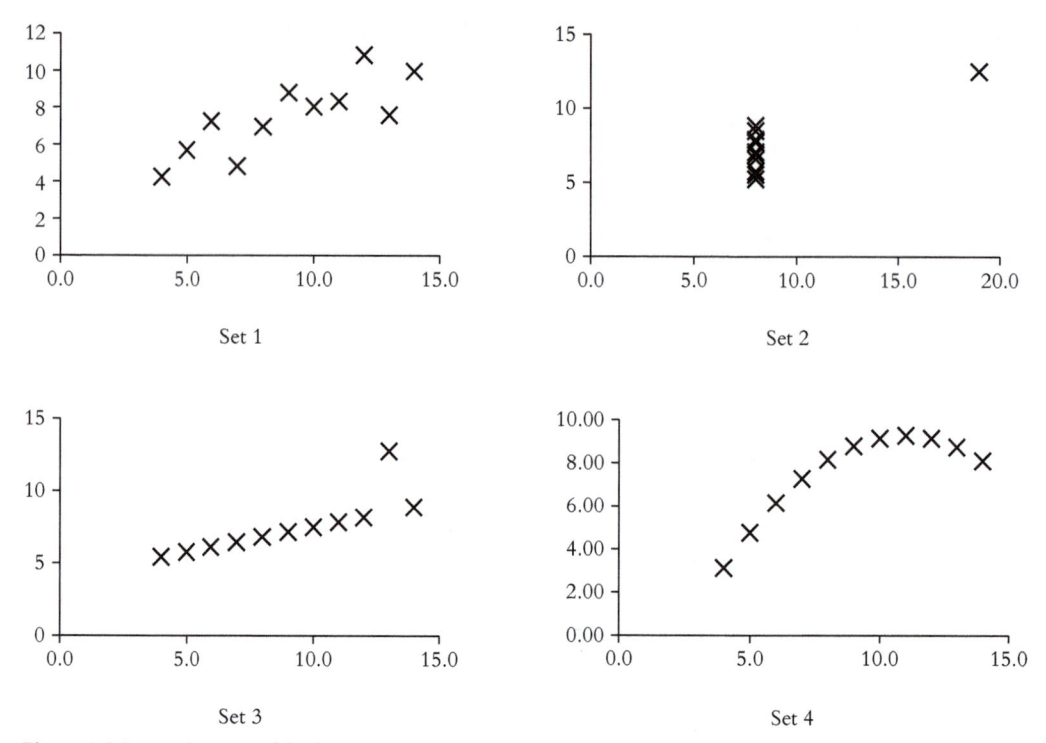

Figure 3.2 Scatter diagrams of the four sets of data in Anscombe's quartet

regression lines) but very, very different distributions as you can see in Figure 3.2.

The obvious message is that the mean and variance do not necessarily tell you much about the distribution of a data set, particularly a bivariate set, and that graphing gives an intuitive and simple way of seeing what is actually going on (see Figure 3.2). Students can be presented with just Anscombe's quartet and asked to find the mean and variance of each data set, perhaps as a follow-up to a lesson on finding and interpreting these values. It becomes clear very quickly what is about to happen, and that these measures are not necessarily trustworthy as a way to distinguish between distributions.

	1		2		3		4	
	x	y	x	y	x	y	x	y
Sample min (Q_0)	4	4.26	8	5.25	4	5.39	4	3.1
Lower quartile (Q_1)	6.5	6.315	8	6.17	6.5	6.25	6.5	6.695
Median (Q_2)	9	7.58	8	7.04	9	7.11	9	8.14
Upper quartile (Q_3)	11.5	8.57	8	8.19	11.5	7.98	11.5	8.95
Sample max (Q_5)	14	10.84	19	12.5	14	12.74	14	9.26

Table 3.3 Five-number summaries for the four sets of data in Anscombe's quartet

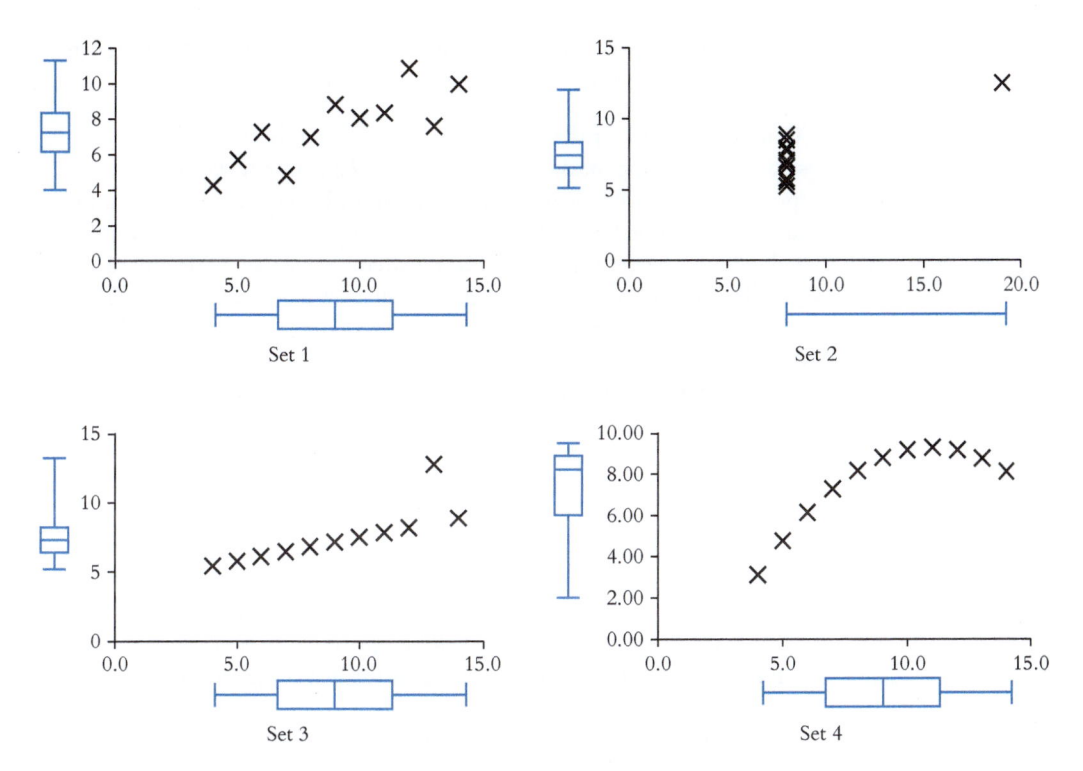

Figure 3.3 Anscombe's quartet with boxplots

So what happens if you use the five-number summary? Table 3.3 shows the values for Anscombe's data sets, followed by the scatter diagrams with boxplots (see Figure 3.3). Each boxplot is simply the visual representation of a five-number summary.

The five-number summaries show some of the differences between the data sets very clearly, because they are not skewed or pulled around by the outliers. Note for example the values for the x-variable in set 2, where it is now clear that at least 75% of the values are 8. Another feature that jumps out is that the five-number summary for x is identical for the other three sets. Does this mean they are the same? As it happens they are, but it is not clear from the scatter diagrams, and it need not be the case.

Note that Anscombe did not create these data sets to say 'look how clever I am to create very different data sets with the same summary statistics', but rather 'look what happens to the mean and variance when you have strange distributions and sampling error, and therefore why it is safer to use robust measures that allow for them, and it is very sensible to try graphing your data to get an intuitive feel for what the distribution is like'. Note also that these measures do not show much difference between the distribution of y-values for sets 2 and 3; it is only in combination with the distribution of x-values that we see that the sets differ, and we need the visualisation to really understand it.

To really get a feel for the robustness, you can take this a step further with students. Try taking the 3rd or 4th data set and add or remove points in a spreadsheet that will calculate the mean, standard deviation and five-number summary, and watch the effect. The robust measures will change very little, and will not change any more with an extreme value than with a more plausible outlier. You can set challenges such as 'change one value to −100 without changing the median or quartiles', or 'change one value to −100 so that the median, goes down', or 'change one value so that it becomes the median', and so on. Students can even set each other challenges.

> ### Data creation classroom challenge
> Why not try getting students to create sets of data to give different types of distribution. For example, create a set of ten values for which the range is as close as possible to 5, the mean is close to 8 and the median is close to 10. Asking for an imprecise answer is deliberate as it is tough to get exact values at a first attempt, and by defining getting near as a success, students have motivation to iterate towards a 'better' answer and to move on to other examples without getting bogged down. Opportunities will also emerge for teachers to explore why, for example, one student's proposed data set has an exact value for the mean but is not close to the median, while another has an exact median but not mean, and to encourage discussion about how the structure of the data set and relationship between individual data points affects the summary values.

3.5 Static data visualisation

Good data visualisation is a careful balance between technical detail and design. The skill is in selecting a good presentation method for the data, not in deciding which standard one to use. This is not to say that standard forms are not useful, but they can become a straightjacket, particularly when easily available as a set of buttons on a toolbar.

Edward Tufte's 1983 book *The Visual Display of Quantitative Information* is a seminal, and very beautiful, work on data presentation. In it, he presents five principles of graphical excellence, which can be summarised as:

- the well-designed presentation of interesting data
- complex ideas communicated with clarity, precision and efficiency
- that which gives to the viewer the greatest number of ideas in the shortest time with the least ink in the smallest space

- being nearly always multivariate
- telling the truth about data.

The first bullet is key; it means that the design of the presentation should match the data and the context. Data presentation is not about choosing the 'correct' graph from a standard list, but about designing a form of presentation that covers these principles. This might use one of the standard forms, possibly with some variation, or it might use something unique or innovative. What students should be developing is an intuitive judgement in choosing tools and representations, with a willingness to try something and decide it was not a good choice. They can then draw informal conclusions from the representations that lead to better representations, to further investigation or to generating hypotheses. To a great extent, the presentation of data is part of the process of exploratory data analysis, not the end point of the process.

Following the example above using Anscombe's quartet, when faced with the four sets of data the initial analysis using summary data shows no differences, so it is reasonable to look for a graphical representation. A scatterplot is an obvious choice. Plotting the sets highlights the (enormous) differences between them, but it does not end there. Some plots show outliers, which in context would be worthy of investigation; some show the suggestion of a linear or non-linear relationship, which again warrants further investigation. Having decided that the classical summary data was not useful, five-number summaries are found, and the graphics are then improved with boxplots, allowing both the marginal and joint distributions to be shown on the same diagram. Even in this small example, the iterative, cyclic and informal nature of the process is obvious.

Tufte's book is full of such innovations, some historical and some his own. One that has made it into the mainstream is the use of 'small multiples' to show the way in which variables within a multivariate data set relate to a given variable. Putting a series of related graphs in a group on a page, using the same axes, and usually keeping one of the variables the same, reinforces the similarities and differences between them. Also, rather than squashing too much data onto a single diagram, the small multiple lets the reader learn to read one chart and then apply that knowledge to the rest, rather than being overwhelmed by the richness of the data.

A classic example from Tufte (1983) is the set of graphs of unemployment and inflation rates over time, which demonstrates the relational and comparative aspect, as well as the use of time as a variable to create a narrative in the graphic (see Figure 3.4). Alternatively, time may often be used as the variable that changes between the graphs, so that the small multiples act like a set of frames or a storyboard. This sort of information could now be represented dynamically by a tool such as *Trendalyzer*, as discussed later, but there is still a value to having the nine countries separated out.

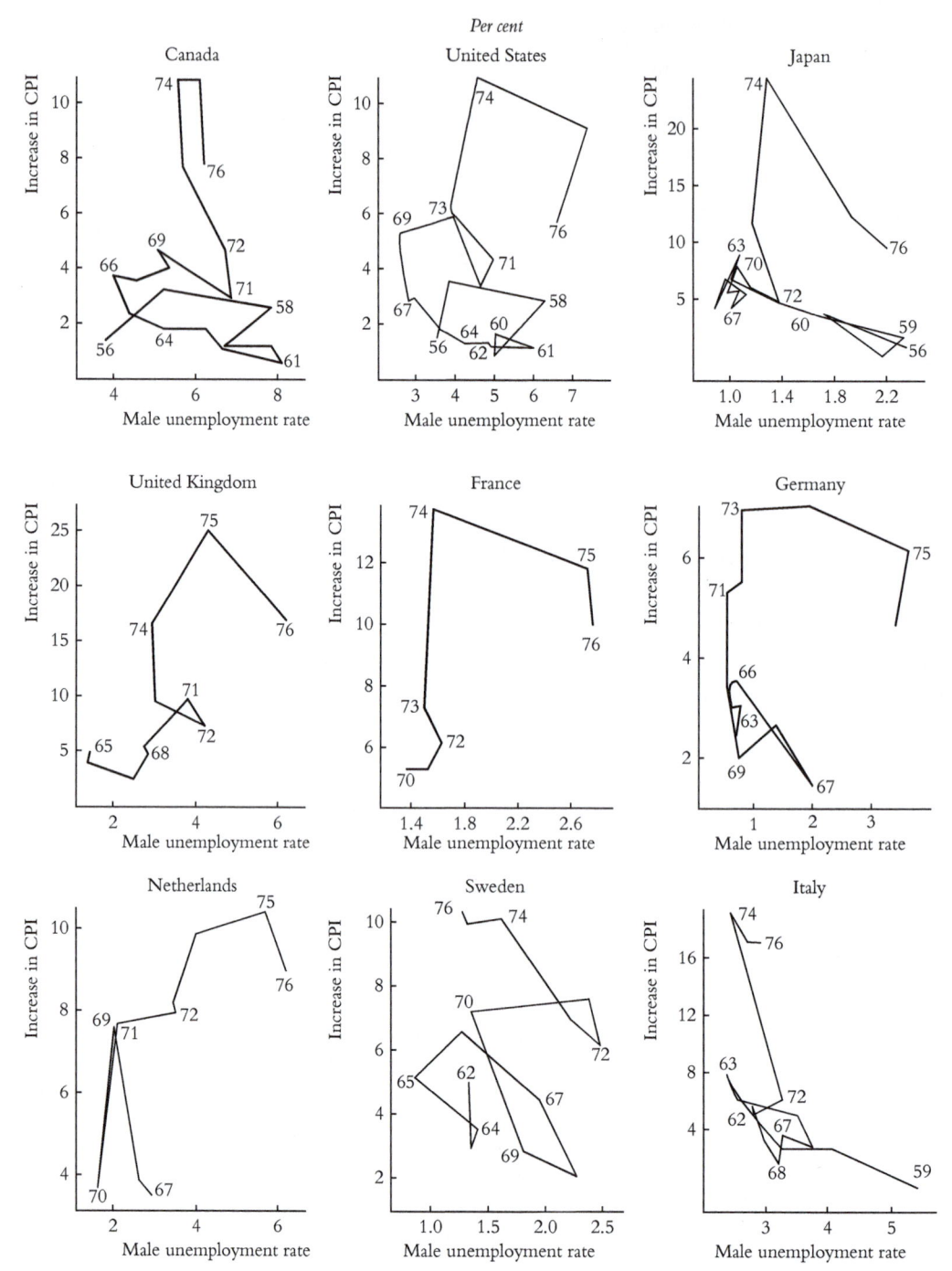

Figure 3.4 Inflation and unemployment rates
[From *The Visual Display of Quantitative Information*, Tufte (1983)]

PART 1 A vision for statistics in schools

A less traditional example is Dan Meth's Trilogy Meter, in which he graphs his enjoyment of the three films in some classic movie trilogies (see Figure 3.5). Once you have worked out what it is you are looking at, this graph is very information rich and it is easy to pick out, for example, that he usually likes the first film, that the second film is surprisingly often the best in a trilogy and that the third film is usually rubbish.

THE TRILOGY METER

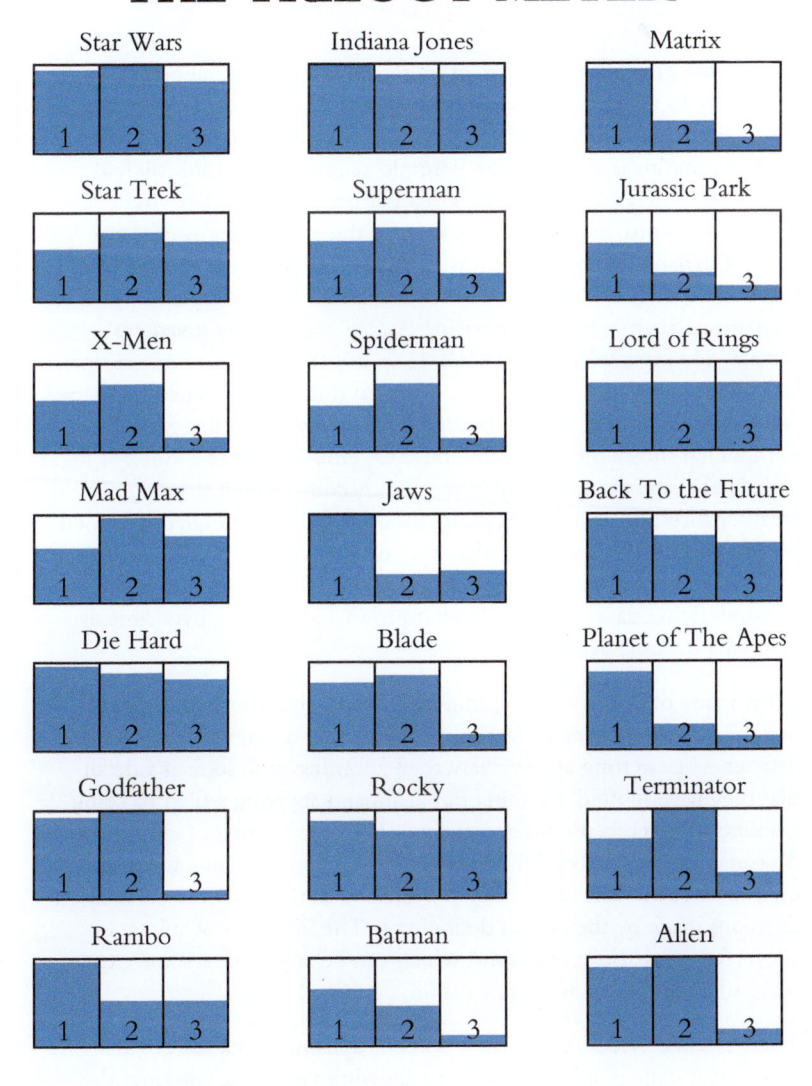

Figure 3.5 The trilogy meter
[From *danmeth.com*]

43

Small multiples like this make great starter activities for lessons on statistics. A few minutes for students to get their heads around what the graphic shows, followed by some discussion of the interpretation, conclusions and/or questions raised would be enough, as well as a critique of the presentation itself and even the method of data collection (who is Dan Meth and why would I want to know what *he* thinks?!).

3.6 Dynamic data visualisation

One of the greatest advantages of increased computational power is the ability to display data dynamically rather than statically. Dynamic data visualisation includes the use of animation and/or interaction to represent multivariate data. For example, technological tools such as *Gapminder* and *Tinkerplots* allow visualisation of multivariate data over extra dimensions, including time. One of the great benefits of these sorts of dynamic representation is that they can allow for an *interactive* experience. Other ways of dynamic visualisation are essentially animations, such as the mesmerising *A Day in the Life of Americans* (http://flowingdata.com/2015/12/15/a-day-in-the-life-of-americans). Dynamic representation can even mean, in this world of open big data, real-time animations such as *Breathing Earth* (www.breathingearth.net), which shows births, deaths and CO_2 emissions in real-time, or the utterly captivating *Windy* (www.windy.com), which shows global weather forecasts. Strictly speaking, these last two are simulations based on data, but this is simply another way of presenting the underlying data set. The other thing to notice about these last two examples is that the underlying data set is not *fixed*; it grows and changes over time as new data become available.

What many of these dynamic, animated, interactive visualisations have in common is that they allow the viewer to take different views on one data set – first getting an overview, then zooming in to look at a detail that they have spotted, zooming out again and applying a filter to bring a different aspect to the surface, and so on. In the context of exploratory data analysis, the utility of this is enormous because it allows for genuine *exploration* of the data set by the viewer, not a static presentation based on decisions made by the person designing it. The inclusion of animation allows for comparison over time, which can be achieved to some extent using small multiples, as we saw earlier.

There is some evidence that these sorts of dynamic visualisations are better than static graphs and tables in allowing viewers to interpret the data, make comparisons, focus on key data and understand change (for example, see Nakakoji, Takashima & Yamamoto 2001), though with the caveat that direct, formal statistical analysis is better done from traditional tables and graphs. This is as you might expect, and supports the use of dynamic visualisation for *exploratory* data analysis.

In the classroom, the main barrier to students creating their own dynamic visualisations is technical skill, but that should not stop students interacting with and exploring other data sets presented dynamically, and *Gapminder* is a great example of an accessible source of very rich visualisations.

Gapminder Foundation (www.gapminder.org) is the brainchild of the late Hans Rosling, a Swedish professor of international health well known for his contributions to TED conferences. The tool which Rosling is famous for using in his lectures is *Trendalyzer*, a dynamic and interactive bubble chart which shows five variables. If you are not aware of his work, then *The best stats you've ever seen* (Rosling 2006) is a good place to start. In the example in Figure 3.6, Trendalyzer is set up to plot life expectancy against income per person for various countries, but it also shows the population of each country as the bubble size, the continent by bubble colour *and* the change over time via the animation.

Figure 3.6 Life expectancy plotted against income per person [From www.gapminder.org]

The wealth of data behind Trendalyzer, the number of different indicators that can be chosen, and the intuitive interface make it an amazing tool for exploratory data analysis in the classroom. The ease of use in particular removes the technological barrier of generating dynamic visualisations, while still leaving room for students to add their own personal choices and design. Following Cobb (2007), it is good to remember that what we are able to do shapes what we think we ought to do, and a tool such as this opens up the number of directions in which we think we *could* explore. As previously mentioned, the use of global data like this can require some bravery in the classroom, and certainly some choices of indicator might be inappropriate for particular groups of students.

There are often concerns when using such tools about the provenance of the data and therefore the strength of any conclusions drawn from it. However, it must be remembered that the idea of exploratory data analysis is to explore, and as the great Hans Rosling once said, when discussing conclusions drawn on a global scale for a very large data set: 'Data is often better than you think. The differences are much bigger than the weakness of the data' (Rosling 2006). Exploratory data analysis helps us to find those differences.

Most formal curricula emphasise confirmatory analysis over exploratory, often ignoring the latter completely. However, introducing students to exploratory data analysis can be extremely liberating and is vital for developing statistical literacy that is applicable in the real world. The willingness to take chances, to go down blind alleys, to spot patterns and to develop formal questions are key skills in learning statistics and support very well the more formal techniques learnt within the curriculum. Encouraging students to think deeply about features of data displayed in graphical representations and how to relate these features to the context of the data helps them develop the ability to think about the data in relation to the underlying structures of the representation, such as the labels on the axes and the type of data shown. According to diSessa (2004), improvements in meta-representational competence may well result in students gaining increased capacity to learn new formal representations alongside the enhancement of more broad-scale problem-solving abilities.

References and further sources

1. Tukey, J. W. (1961). The Future of Data Analysis. *The Annals of Mathematical Statistics*, 33(1), 1–67 (out of print but freely available online). <https://projecteuclid.org/download/pdf_1/euclid.aoms/1177704711> accessed 6th November 2017.

2. Tukey, J. W. (1977). *Exploratory Data Analysis* (Addison-Wesley Publishing Company).

3. Moore, D. S. (1992). Teaching Statistics as a Respectable Subject. *Statistics for the Twenty-First Century.* Eds Florence Gordon and Sheldon Gordon, MAA Notes, No. 26 (Mathematical Association of America), 14–25.

4. Cobb, G. W. (2007). The introductory statistics course: A Ptolemaic curriculum? *Technology Innovations in Statistics Education*, 1(1), 1–15.

5. Tukey, J. W. (1980). We need both exploratory and confirmatory, *The American Statistician*, 34(1), 23–25.

6. Ansombe, F. (1973). Graphs in statistical analysis, *The American Statistician,* 27(1), 17–21.

7. Tufte, E. (1983). *The Visual Display of Quantitative Information* (2nd edn, 2001, Graphics Press).

8. Nakakoji, K., Takashima, A. & Yamamoto, Y. (2001). Cognitive effects of animated visualization in exploratory visual data analysis. *Proceedings of the Fifth International Conference on Information Visualisation* (Washington, DC: IEEE), 77–84.

9. Rosling, H. (2006). The best stats you've ever seen. TED talk, <https://www.ted.com/talks/hans_rosling_shows_the_best_stats_you_ve_ever_seen> accessed 6th November 2017.

10. diSessa, A. A. (2004). Metarepresentation: Native Competence and Targets for Instruction. *Cognition and Instruction*, 22(3), 293–331.

Chapter 4

Simulation

4.1 Introduction

The word 'simulation' conjures up visions of various complex systems which attempt to model the real world, such as systems for testing the emergency procedures in a hospital, flight simulators, 'sandbox' computer games like *SimCity* or virtual surgery for trainee doctors. These are very complex simulations designed to replicate (to some degree of accuracy) a real-life situation in order to try out systems, practise skills or play games. Simulation is also used to model physical systems, as opposed to the examples above which have human agents, in order to gain insight into the way that they function and interact with other systems.

The simulations we are going to look at in this chapter are much, much simpler, but they have a similar function; they are designed to replicate the outcomes of a situation where it is difficult, impossible, dangerous or tedious to gather enough actual data to see what will happen in the long term. These situations need not be physical systems, or even real-world systems; we can also use simulation as a way for students to estimate probabilities using empirical data. Using either physical simulations or computer-based simulations, students can discover or confirm the law of large numbers, which says that the more data they collect, the closer their empirical distribution gets to the theoretical distribution. More practically, students can also generate empirical distributions for experiments for which they don't currently know enough probability theory to construct theoretical distributions. They can then use these empirical distributions, if informally, to test hypotheses. In addition to this modelling of complex systems, we can also use simulation in teaching statistics to mimic the process of taking samples. This is a very useful pedagogical tool to stimulate thinking about variability in sampling and the effects of changing sample size or sampling technique, as well as to demonstrate more formal techniques such as bootstrapping in which you take multiple subsamples from a sample and use them to infer the population parameters.

The use of simulation in such ways, even very early in the curriculum, means that students can be exposed to some deep statistical ideas without having to wait until they have learned the technical skills needed for the formal approach. Currently, some students never meet these ideas since they stop studying statistics before reaching the requisite level of technical competence. This approach is outlined in the GAISE pre-K–12 framework (Franklin 2007), and some examples of it can be found in the activities that accompany this chapter (see chapter 11).

While formal inference in the form of hypothesis testing is not prevalent in secondary curricula, the ideas are valuable when introduced as part of students' exploration of probability. It is not uncommon for students to examine probability in terms of equally likely outcomes using combinatorics and sample spaces, but this potentially limits their studies to rolling dice, drawing marbles from bags and flipping coins. Opening up the exploration of probability as a process of modelling physical systems statistically allows creative approaches to lesson design that support and supplement the theoretical probability students are expected to learn. It may also help students to see theoretical probabilities as expectation values within a variable process that stabilises in the long term, rather than as a definitive predictor of individual trials. Using simulation to generate empirical distributions means that we introduce distributions as possible models for the patterns observed in the data or simulation, not as abstract probability distributions. Given some data from a real-world system, we can hypothesise something about the way in which we think the data might be generated, then simulate it and compare the results with the data to decide how good our model is. Later on, when students study more formal techniques, this experience will help them to keep in mind that the formal distributions are ultimately models for things, not true reflections of those things, and to consider the effect that this has on the inference that they draw. This might be something as simple as using probabilities from rainfall forecasts to make a spinner for simulating the rainfall over the next week, and then deciding whether the actual rainfall was as expected.

The word 'simulation' does tend to imply that some form of technology is involved, but as suggested above, this is not necessarily the case. Some of the examples that we consider involve students using physical random number generators such as spinners or, when lots of random numbers are needed, harnessing technology to generate a long sequence. While it is tempting to use technology to generate large amounts of data, the physical process of (slowly) generating data allows students to develop an explicit understanding of where the short-term variation and long-term stability comes from. With technology this can often be hidden 'under the hood'. It is not always immediately obvious to students that a physical experiment and the computer model are really doing the same thing. Following an experiment that involves physically throwing dice with a computer-based experiment that models throwing dice, as suggested by Nicholson (1997), can reassure students that the computer produces similar results. This can help them to trust the outcomes of computer-based simulations that they cannot replicate physically. Another advantage of doing concrete simulations with manipulatives is to get a feel for how the system works before scaling up with technology to get a mountain of data. It is also an opportunity to explore ideas about sources of error such as measurement or recording errors, since once the process is automated, these won't occur and outliers become an indication of variability rather than a possible error.

It is important to get students to think about how random their numbers are: what is the difference between numbers generated by rolling a die and a published list of random numbers or a set of numbers output from a computer? There is no need to delve too far into the idea of pseudorandomness, but students do need to understand, for example, why it is not acceptable to use the same string of numbers as a seed for multiple runs of a simulation. See the box 'How random is random?' in the next section.

Finally, there is a tricky balance to be struck in delivery using simulations. On the one hand, it is very tempting not to 'waste time' in allowing students to conduct simulations for themselves, leading to a demonstration from the front and highly passive learning. On the other hand, simply throwing a complex simulation experiment at students and expecting them to extract the important principles, or happen to try useful avenues, is also unrealistic. The middle ground is to give students some questions to answer before they perform the simulation or to consider as they carry out the simulation. This allows the teacher to guide the students along the desired pathway and to steer them towards thinking about the particular aspect that is the focus of the lesson. See Lane & Peres (2006) for a good discussion of this and some further reading.

4.2 Some practical considerations

Simulations, particularly those run by hand, can often go 'bad'. The random nature of the input does mean that some students will sometimes get a very odd result simply due to random fluctuations. This is something that needs to be drawn out in interpretation and conclusion of the activity. In activities where each student, or group of students, is generating numbers using tools such as spinners, you should always include either whole-class discussion of outcomes or aggregation of data in order to mitigate such problems. One of the many benefits of computer-based simulations is the ease with which multiple simulations can be run. This allows each student, or group of students, to run many instances so that they can see the variability for themselves. If they then get a particularly unlikely outcome, then they see it in context. Chapter 5 introduces the idea of a 'growing samples' activity, which can be a good starting point for helping students to think about the number of runs necessary for a simulation to produce useful results.

There are many ways to source 'random' numbers with which to run simulations, but they have different statuses. You probably don't want to spend too much time in the classroom discussing the difference between quasirandom, pseudorandom and true random numbers, but it is an important consideration.

How random is random?

True random numbers can be generated by physical processes such as rolling dice, using spinners and so on, but it is a very slow process if you want a large quantity of numbers. There are various issues with measuring physical phenomena to obtain random numbers, but www.random.org is a good source of free, true, random numbers based on atmospheric noise if you would like a large set and don't have access to a published table of numbers.

Pseudorandom numbers are generated in sequences by an algorithm based on an initial 'seed' value; while they approximate a set of random numbers, they are ultimately deterministic since the algorithm will produce the same sequence of numbers from the same initial state. However, for almost all practical purposes they are 'random enough'. The random numbers generated by a spreadsheet program will be pseudorandom, and this is what you will be using when simulating using technology. In most cases the seed is chosen by the computer based on a value derived from information available to it such as the current time, but in some software, the seed can be selected by the user in order to generate the same sequence of pseudorandom numbers in the future. For teachers wishing to prepare material in advance, this can be a useful way of maintaining some control of an apparently random process. In the statistical software R, choosing a seed value can be done using the `set.seed()` command before generating random numbers. To recreate a specific string of random numbers is easy: next time the 'random' numbers are generated, use the same seed value inside the parentheses as before.

Quasirandom numbers are sequences of numbers where the proportion of numbers in any interval is roughly the same as the proportion that interval is of the whole range of possible values. This has the advantage that they cover the whole range of possible values very quickly and in the same sort of proportion as you would expect from long-term behaviour. In a context in which it takes a (very) long time for long-term behaviour to settle down enough to draw conclusions, these numbers can be used to speed up the simulation. If you carry out the law of large numbers activity (Activity 2 in Chapter 11), you might like to follow it with the same activity using quasi-random numbers to show how quickly it converges. You can source sequences from various places, including here: http://qrand.gel.ulaval.ca/data/uploads/dr12-ghs.json

4.3 The law of large numbers

When taking small samples from a population, we would expect to see large amounts of variation between the samples, but as the samples get larger, we would expect the samples to more closely resemble the population. Put another way, as you increase the size of a sample, the tendency is for the parameters of the sample to tend towards those of the population, as will the proportions of subpopulations. In the context of sampling, this is how it is possible to infer things about a population from a decent-sized sample or set of samples. In simulation, it is the powerful effect that allows us to use technology to work out difficult probabilities and investigate the long-term behaviour of complex systems.

In fact, even some very simple systems can be tedious or time-consuming to investigate physically. For example, consider the situation where you flip a single coin multiple times, keeping track of the total number of heads. How would you expect the proportion of heads to vary as you continue to flip the coin?

Over time you would expect the proportion p to tend towards 0.5, but would it do so monotonically (start at 1 or 0 then head towards 0.5 without going the 'wrong' way)? Would it get within 1% of 0.5 very quickly, or would it take a long time? The answers to these questions are quite counter-intuitive, and are very time-consuming to explore with a coin or spinner. Even with 1000 coin flips you still end up with $p < 0.48$ or $p > 0.52$ on occasion. However, these questions are very easy to explore with technology and a simulation, as in Activity 2 in Chapter 11. In this activity, pressing F9 will generate a new set of 1500 values along with the graph showing the journey that the sample average has taken. Good, or interesting, examples can be kept visible by copying and pasting each one as a picture.

This can be a good exercise in training intuition. Start by getting students to do a small number of coin flips by hand, say 20, and calculate the sample average for their sample after each flip. If they can plot a graph of their sample averages against the number of flips, so much the better. They can then compare graphs. Whose graph shows the 'true' value? Whose shows strange or unusual behaviour? How unusual is it within the class?

Students can then move on to the spreadsheet, having first discussed and predicted what will happen when you introduce a large number of data points. Intuitively, one expects the line to stay very close to 0.5 once it has got there, but this is often not the case and it may drift away from the middle. This contrast between long-term stability and short-term variability is a key concept in general statistical literacy, so it is worth drawing out the idea through this or similar activities (Reading & Reid 2004).

While this particular activity is aimed specifically at thinking about the law of large numbers, this concept is one that necessarily underlies all of the other simulation activities as well.

4.4 Empirical distributions

'I flipped this coin 10 times and got 8 heads. Does this mean it is biased?' This simple question about variability is tricky to answer directly, and of course any conclusions drawn about the fairness or otherwise of the coin can only be inferential rather than deductive.

To decide whether or not you have a biased coin you have to make a comparison to something solid. The standard procedure is to assume (hypothesise) that the coin is fair, because this is something you can model concretely, then look at the distributions of outcomes you would expect from that and see how unlikely the observation is. You then have to decide how unlikely you want an observation to be to judge that the original assumption is false and you have a biased coin.

This is a hypothesis test. We are not going to discuss the formalities of hypothesis tests, but the concept of comparing data against a theoretical distribution is perfectly possible for students to grasp before they meet all of the theoretical underpinnings in post-16 mathematics or statistics courses, particularly using simulation. One use of simulation is to generate the distribution of outcomes, particularly in a case where working it out theoretically would be difficult or impossible. The distribution of outcomes when you flip 10 coins is binomial, so we could carry out this test using the binomial distribution, but in fact we can do pretty well by simply randomising outcomes to see what sorts of results are usual and what are unusual. Crucially, there is no need to be able to calculate binomial probabilities to understand how this works, because you can compare your sample against the simulated distribution rather than the abstract one. One major pedagogical benefit of this is that the model we are using and the method used to generate the data are very closely related, so it is easy to see the link between the data, the model and the inference we make based on that model (Cobb 2007).

A simple activity based on this is to use a spreadsheet to simulate a large number of sequences of 10 coin outcomes. It is very simple to simulate 1000 or 10 000 data points, by which time, following the law of large numbers, you have something very close to the theoretical distribution. The key is that we don't need the accuracy of the theoretical distribution; the empirical one is good enough so long as the value of n is big enough.

Returning to the original question, we should find that the probability of 8 or more heads is about 5 or 6%. Note that we don't really care what the precise value is; the empirical distribution tells us that about 1 in 20 trials with a fair coin would come out with as many, or more, heads as we got. This doesn't seem very unlikely, so we could easily interpret our 8 heads out of 10 as a 'blip'.

We have to be very careful about the conclusions that we reach from such a test. We have not *shown* the coin to be fair; it is simply that the evidence we have is not strong enough to allow us to infer that the coin could be biased. We should therefore conclude with something like 'There is insufficient evidence to suggest that the coin is biased.'

This is a fairly simply context, but it is easy to see how the ideas can be applied to more complex situations in which the distributions are less easy to find analytically.

Another good example which is distribution free, i.e. where the distribution from which the sample data comes is unknown, is given in Ernst (2004), but is presented here with a simulation spin given to it by Cobb (2007). The post-surgery recovery times, in days, of seven patients are as follows, where the first three patients were given the standard care and the next four were given a new treatment: 22, 33, 40, then 19, 22, 25, 26. The question is whether or not this data provides evidence that the new treatment reduces recovery time. You can present the raw data like this to your students and simply ask for an intuitive answer; alternatively, encourage them to calculate a statistic of some sort, for example the mean recovery time for each group. For the first group it is 31.67 days and for the second 23.00 days. Is this difference of 8.67 days enough to be considered significant? Does it imply that recovery times have reduced, or is the result within the bounds of natural variation?

The issue here of course is that we don't have access to the distributions of recovery times, nor to very large numbers of patients to keep repeating the trial. What we can do is to set up what is called a permutation test.

Assume that the treatment does nothing at all. As in the previous example, this gives us a baseline model to compare our data with and decide if it looks 'weird' enough to suggest that our model is wrong. The beauty of this is that, under this assumption, the only difference between the patients is the group to which they were assigned. If the treatment didn't do anything then the times above would be the same with or without it. All we have to do then is to simulate assigning the seven patients to the two groups multiple times and see what the difference is in the mean recovery times.

It turns out that a difference of means of 8.67 or more happens about 8.6% of time (it is chance that these two numbers are very similar), which is not common, but also not unusual enough to get excited about; so we conclude that there is insufficient evidence to suggest that our model is wrong, i.e. there is insufficient evidence to support the claim that the treatment reduces recovery times.

Note that, because we start by assuming that the treatment does nothing, all we are left with is the *permutation* of patients. In this particular case there are only 35 permutations, so we could probably list them all and work out this distribution, but it is easy to see that in some contexts combinatorial explosion will lead to an unwieldy number of permutations, leaving simulation as the most efficient method.

While these ideas go beyond the expectation of many pre-16 curricula, if we value statistical literacy as an entitlement of all students, it is necessary to explore beyond simply comparing the means of two groups, and simulation gives a 'mathematics-lite' option. The importance of an engaging context in helping students access advanced ideas cannot be overestimated, and, should medical interventions not prove stimulating to particular groups of students, it is relatively straightforward to create small-scale trials in a classroom that students can take part in.

> **Classroom trials**
>
> Question: Is it easier to identify a branded soft drink if the drinks are chilled?
>
> Select a popular branded soft drink and a cheaper version. Split students into two groups, A and B.
>
> Group A: Do a taste test on the drinks at room temperature; each student scores 1 point for guessing correctly and 0 points for guessing incorrectly.
>
> Group B: Repeat the test for the same drinks but this time chilled in a fridge, or with ice added.
>
> Find the mean score for each group and compare the means. Is the difference significant?
>
> To confirm, feed the data for all individual students into a permutation simulation.

4.5 Modelling

In Part 2 of this book we give details of an activity in which simulation is useful for modelling the experience of collecting toys. We avoid the possible technical difficulties of implementing this scenario with technology, by using manipulatives. This has the dual benefit of allowing students to develop a really good intuitive feel for what is going on in the context and of making the activities properly accessible even for fairly young students. The disadvantage of producing less data can be avoided by collating data from across the class.

There are many other situations, such as queuing or the spread of disease, which can be simulated with pen and paper, many of which include complications such as human agents making decisions. We will not discuss these further here, other than to stress the importance of remembering that what the students are doing is modelling. Underlying assumptions about probabilities, behaviours of human agents and so on should be discussed – not simply 'what assumptions have I made?', but 'what are the implications of this?', 'what different assumptions *could* I have made?' and 'how do my assumptions affect the validity of my conclusions and the appropriateness of my model?' A good example (Selkirk 1973) is simulating football league tables based on nothing but random numbers and some basic proportional assumptions. It produces a league table in which, although the specific teams don't necessarily match the previous year's rankings well, the distribution of points to position in the table is pretty close. Of course this particular context is likely to be singularly unappealing or even off-putting to many students, but the basic principle can be adapted to create more interesting scenarios for individual classes.

By using ideas of probability and simulation to generate models of statistical phenomena, students are able to have a much deeper experience of both, laying the groundwork for both higher-level study and critical evaluation of claims made in the real world using more advanced techniques.

References and further sources

1. Franklin, C. A. (Ed.) (2007). *Guidelines for Assessment and Instruction in Statistics Education (GAISE) Report: A Pre-K–12 Curriculum Framework* (Alexandria, VA: American Statistical Association).

2. Nicholson, J. (1997). Developing Probabilistic and Statistical Reasoning at the Secondary Level Through the Use of Technology. *Proceedings of the 1996 IASE Round Table Conference*. Eds J. Garfield and G. Burrill (Voorburg, The Netherlands: International Statistical Institute).

3. Lane, D. M. & Peres, S. C. (2006). Interactive simulations in the teaching of statistics: Promise and pitfalls. *Proceedings of the Seventh International Conference on Teaching Statistics* [CD-ROM]. Eds A. Rossman and B. Chance (Voorburg, The Netherlands: International Statistical Institute).

4   Reading, C. & Reid, J. (2004). Consideration of variation: A model for curriculum development. *Curricular Development in Statistics Education, Sweden, 2004* (International Association for Statistical Education), 36–53.

5   Cobb, G. W. (2007). The Introductory Statistics Course: A Ptolemaic Curriculum? *Technology Innovations in Statistics Education*, 1(1), 1–15.

6   Ernst, M. D. (2004). Permutation Methods: A Basis for Exact Inference. *Statistical Science*, 19(4), 676–685.

7   Selkirk, K. (1973). Random Models in the Classroom 1: An Example. *Mathematics in School*, 2(6), 5–6.

A simulation tool with good visualisations and intuitive interface can be found at www.lock5stat.com/StatKey

Chapter 5

Sampling and variation

'Statistics is a general intellectual method that applies wherever data, variation, and chance appear. It is a fundamental method because data, variation, and chance are omnipresent in modern life. It is an independent discipline with its own core ideas rather than, for example, a branch of mathematics.'

Moore 1998

5.1 Introduction

The difference between statistics and mathematics boils down to the non-deterministic nature of the data generated by a statistical process. The basis of all statistical investigations lies in a fundamental feature of anything that can be measured, variation. Students' early experiences of working with data should therefore involve recognising the ever-presence of variability, developing the ability to recognise variability, and explaining why variability occurs. When planning experiments, students must develop strategies to minimise variation where possible.

It can easily be argued that understanding and explaining the nature of variability in a given context is essential if a statistician wants to draw conclusions that have any hope of being valid. For all people who come into contact with data from politicians, news media, sales people, or any other potentially biased source, a broad understanding of variability can provide an easy sense-check of claims based on data. It is often the case that conclusions are put forward as definitive without any recognition of the role of the underlying variability, and spotting this is key to identifying which sources of information are reliable.

In 2016, polling organisations received much public criticism for 'incorrectly' predicting the results of both the British referendum on membership of the European Union and the election of the next president of the United States of America. In reality though, while many media reports portrayed the poll-based predictions of these close-run elections as definitive, the polling organisations themselves gave their predictions within margins of error that could have seen the results go the other way. The polling organisations recognised the impact of variation and potential bias in their samples, but the media did not report it.

After being neglected in many curricula, the status of variation as an essential component of statistical study is on the rise. Education research now includes the exploration of variability as a fundamental component of statistical literacy (Reading & Reid 2004). Variation also provides a justification for the importance of sampling, along with a way of understanding how and why sampling sometimes works as a representation of a wider population. Exploring variation provides an opportunity to examine the link between probability and statistics in an informal way before students encounter formal inferential techniques during higher-level study.

> **The core elements of statistical thinking**
>
> The core elements of statistical thinking can be described as follows (National Research Council 1990):
>
> 1. The omnipresence of variation in processes. Individuals are variable; repeated measurements on the same individual are variable. The domain of a strict determinism in nature and in human affairs is quite circumscribed.
>
> 2. The need for data about processes. Statistics is steadfastly empirical rather than speculative. Looking at the data has first priority.
>
> 3. The design of data production with variation in mind. Aware of sources of uncontrolled variation, we avoid self-selected samples and insist on comparison in experimental studies. And we introduce planned variation into data production by use of randomization.
>
> 4. The quantification of variation. Random variation is described mathematically by probability.
>
> 5. The explanation of variation. Statistical analysis seeks the systematic effects behind the random variability of individuals and measurements.

Fundamentally, to 'do' statistics, we must appreciate that some processes are non-deterministic in nature and require data to be collected. During that process of collection, variation must be anticipated, planned for, quantified and explained.

Even though key measures of variability such as standard deviation are generally not encountered by students until higher-level study, some of the more informal concepts can and should be introduced during students' initial experiences with data. In many academic papers the most straightforward recommendation for teaching is to include discussion about the variability in data as a key part of the teaching and learning process (Torok & Watson 2000). Asking probing questions about students' expectations of data can help to move them from focusing on individual values and measures of centre towards a view of statistics that considers the aggregate properties of the system.

A classic task that appears frequently in educational research involves taking coloured lollies from a bag. Students are told that a bag contains 100 lollies of various colours in given quantities such as 50 red, 20 blue, 15 yellow and 15 green. Students are asked to predict how many red lollies will be drawn in each of 6 handfuls containing 10 lollies. How each student answers, particularly the amount of variation in their predicted samples, can give clues as to their underlying thought processes. A student who answers 5, 5, 5, 5, 5, 5 is likely to be treating the process as deterministic, reflecting the proportions in the bag exactly with no attention to the variability inherent in statistical processes. Other students will overestimate the number of red lollies, focusing simply on the fact that red is the most common, rather than reflecting a specific proportion of the population.

In classrooms, this task can be easily adapted to promote development of ideas by first asking students to write down and graph their predictions for drawing several handfuls of lollies, and then asking them to try the experiment (using counters or coloured cubes if the prospect of a classroom of sugar-fuelled students is too terrifying a scenario to contemplate). The process of asking students to make a prediction, along with a justification, before carrying out an experiment is a useful technique for developing statistical reasoning skills, though you should also make sure that students have the opportunity to review their predictions in light of the outcome of the experiment.

An added advantage of this kind of activity is that it is a first opportunity for students to experience the idea of a sampling distribution, in which a sample is taken repeatedly and a statistic is calculated for each sample and used as the basis for a graphical representation. These sampling distributions are fundamental to more advanced ideas of statistical tests, inference and the central limit theorem, but need not be hidden from students until they meet concepts of formal inference.

Research around various versions of the lollies task suggests that before students can fully appreciate the nature of variability in statistics, they must have a strong sense of proportional reasoning and be confident in using percentages and fractions. Many curricula include the production and interpretation of cumulative frequency diagrams and boxplots, but while students can usually be trained to construct these graphical representations and read key values such as median and quartiles, they rarely explore in detail how these values help to locate and describe the aggregate data.

In order to help students develop an idea of how data is located within a graph, opportunities should be provided to help them connect their

PART 1 A vision for statistics in schools

ideas of proportion to the aggregates of data in a graphical representation. One approach is to focus on the representation of frequency by area, an idea that is covered in histograms as part of the standard recipe to produce such diagrams. Asking students to construct diagrams of discrete, quantitative data where each individual data point is represented by a unit square can help create schema for the location of data in symmetrical and skewed distributions.

Figure 5.1 shows a data set of size 100 with the deciles coloured in. You can see very easily that the value 9 has a frequency equivalent to two deciles, but the first decile covers four different values. It is also easy to see the skew from the relative width of the deciles on either side of the mode. In this case the ones below the mode are a bit splodgier and the ones above the mode tend to be taller and thinner.

There are rich links to be made with mechanics for those brave enough to spend some time exploring a geometric approach to distributions. Focusing on locating the balance point of distributions by creating 2D models of graphs using card can help demonstrate a broader idea of the

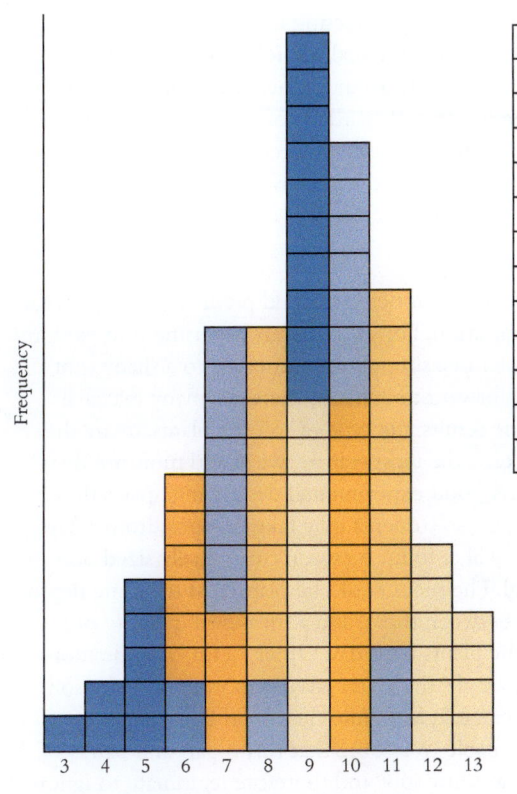

Number of leaves	Frequency
3	1
4	2
5	5
6	8
7	12
8	12
9	20
10	17
11	13
12	6
13	4
total	100

Figure 5.1 Data set with deciles coloured

mean and could be an invaluable early experience for students going on to further study in physics or engineering-related disciplines.

As students develop a sense of the geometric distribution of data in a graphical representation, they will have a richer set of tools to use when critiquing and comparing data presented using boxplots or five-number summaries.

5.2 Types of variability

In order to draw useful conclusions from data, students must be able to identify potential underlying causes of variation and design methods of data collection to mitigate them, or explain the variation where it appears in an experiment.

Variation caused by external factors

No statistical experiment exists entirely in isolation from external events, although some of these external factors may be so complex as to appear the result of random processes. Any external factors that can be anticipated should be controlled for. Einstein famously exclaimed 'God does not play dice with the universe', rejecting the idea in quantum mechanics that interactions of fundamental particles were entirely random in the belief that sufficiently accurate measurements would allow these interactions to be predictable. Regardless of whether this eventually proves true or not, quantum physics has been incredibly successful at using statistical techniques to make the astonishingly accurate predictions necessary to run microprocessors in computers and modern telecommunications.

Students should become adept at anticipating and predicting the external factors that might cause variation. For example, an experiment to explore how tall a plant grows when in a sunny spot compared to a shady spot may initially seem straightforward to run, but there are many external factors that could affect the results: the type of soil the plants are in, the amount of water they receive, the temperature, wind and moisture they are exposed to and so on. A good experimental design anticipates these factors and controls for them, so students may take compost from a bag, mix it well and share an equal amount between two equally sized pots made of the same material. The seeds could be planted at the same depth with the same separation between them, and a measured quantity of water could be given to the plants each day. Other factors may be more difficult to control. For example, the water may evaporate from the soil faster for the plant in direct sunlight. While this could be a source of variation, it might be reasonable to assume that this is part and parcel of the process of growing in a sunny spot and therefore legitimate to ignore as a fundamental part of the system being tested. A good experimental design might collect enough data to see the variability – for example, growing 10 plants in a shady spot and 10 in a sunny spot so that variability *within* a given growing condition can be compared against variability *between* conditions, to account for the aspects which cannot be controlled for such as the quality of individual seeds.

Variation caused by measurement

The bottom line with any statistical process is that things have to be measured or counted, and this can be fraught with problems. This type of variation can be demonstrated very easily to students by giving them an identical angle or length and asking them to measure it individually. Collecting the data and graphing the results will clearly demonstrate the range of answers that can be produced despite the task being straightforward. Again, a virtue could be made of this kind of measurement error when teaching unrelated skills. In a lesson where students are learning to use a protractor, ask them to keep a note of the differences between their answers and the correct answers. After some practice, repeat the exercise and get them to compare the average errors before and after. In this way, students' attention can be drawn to both the value of practising key skills and the prevalence of measurement errors. Such 'measurement error' can be extended to questionnaire responses, where imprecise language may cause two people to interpret a statement in very different ways or assign the same attitude to different values on a response scale. Students should be given opportunities to experience how measurement errors can arise and devise strategies to reduce them.

To explore this process, ask students to measure the length of an object such as a classroom desk using only their hand as the unit, without conferring with each other, and then collect their answers and graph the results using a bar chart or dotplot. It should be a fairly messy distribution as students will more than likely have used a variety of different approaches – spread hands, fingers together, heel of the hand to fingertips and so on – along with different decisions around accuracy. Did they measure to the nearest whole hand? Did they round up or down?

After a brief discussion of how this could be resolved by, for example, agreeing to use hand span from the tip of the little finger to the tip of the thumb on a spread hand, ask the students to measure again and once more make a graph of the results. This should produce a graph that looks more consistent, although there will still be some spread in the data. Further conversations could be had about the different hand sizes of each student and the activity could go on to explore what a 'representative' hand span would be for the whole class. Students could then create a unit measure from this that could be used to measure again to further reduce the variability of the results. It is important to note that while teacher preference may be that this unit measure is based on an average hand for the class, students may come up with valid reasons to use something different. For example, the smallest hand might give more accurate measurement if they choose to only measure to the nearest whole unit. Allowing students to engage in this kind of reasoning is an important part of developing statistical literacy.

Variation caused by error

While a more minor cause of variation, the effect of errors in data collection or recording should be kept in mind by students when exploring data. It is common for students to learn formulae for identifying outliers, such as 'upper quartile $+$ (1.5 $\times$ interquartile range)', and most students are capable of identifying outliers using these techniques, but the decision as to whether to exclude an outlier once identified should be influenced significantly by the context. In data on the weights of newborn babies, it might be reasonable to consider that a newborn appearing to weigh in at 10 kg, more than three times the average birth weight, is likely to be the result of an error in recording data. However, for data on the income of people in a company, the salary of the Chief Executive Officer could easily be more than 20 times that of an average employee, but still be considered an accurate value. It might be valid to exclude their salary (and the salaries of other senior executives) when trying to calculate a representative value for employee income, due to the weight of the fat tail. Alternatively, the CEO's salary might be left in if a resistant measure of centre (such as the median) is chosen. Sometimes the decision about whether to remove the outlier depends also on what you are doing with the data. Students should be given the opportunity to explore data and make decisions on whether surprising values can be explained within the context of the data and whether such values should be removed as part of the process of cleaning the data.

Variation caused by variables of interest

Any complex system is affected by the interactions of a significant number of independent and dependent parameters, all of which have effects on the individual variability, although different combinations may have more or less significant effects. In the era of big data, extracting information from a large system about the effects of variation caused by certain parameters has the potential to provide incredibly rich information.

Studies reported in the *New York Times* have used data collected from Google searches to identify side effects caused by interactions between drugs. By looking at the association between searches for two different drugs and their side effects, scientists are able to identify side effects that appear to occur only when someone searches for both drugs within a short time frame, suggesting that the person is currently receiving both. The complex interactions are identifiable by statistical techniques but would be impossible to anticipate in clinical trials. On a more classroom-oriented scale, students can be prompted to consider the effect on variability of far less complex parameters, for example discussing the exam results in an ability-grouped maths class and in a mixed-ability class. Alternatively, they could discuss what might happen to the variability of the final grades of students in a year group if the school policy was to focus on just the weakest students, or just the most able students.

Natural variation

In the hand span task mentioned previously, students were asked to come up with a representative hand span for their class. Some time could be spent during this kind of task focusing on the natural variation among students in the class and the sources of that variation such as age, gender, genetic factors and so on. Drawing attention to the different types of variation, some of which can be minimised and some of which must be accepted and accounted for, is an important part of the process of developing statistical thinking. Again, opportunities to explore data through different lenses should be encouraged. While a frequency plot of different hand spans would be the first graph considered by many students and teachers, a case value plot of hand spans sorted by date of birth might show other interesting patterns in the data and explain the most common values – for example, are there lots of students with a certain hand span because, by chance, there are a large number of students in the class born close to each other?

Random variation

Random variation is a fundamental aspect of stochastic analysis and has the potential both to obscure the underlying information contained within the data and to reveal patterns that could not be seen by individual observations. The key for students is to gain an understanding of how probability works within statistics. Any individual observation contains little information about the entire population as it could have come from anywhere within the body of possible data points, sometimes in the tail, sometimes close to the centre. Long-term stability and the law of large numbers ensure that given a sufficiently large sample of data, the proportions of subpopulations and the overall shape of the data will eventually reflect that of the overall population. It is precisely this feature that allows a population parameter to be inferred from sample statistics, and although the relationship between the size of the sample and the accuracy of any estimate is complex (and technical), the essential idea that a larger sample produces a more accurate picture of the population is simple to grasp. The role of random variation should not be underestimated in complex systems; interactions between dynamic elements may be deterministic if the precise nature of every interaction could be measured and quantified and a large enough computer brought to bear, but in reality no statistician will ever have every facet of every detail of every initial condition mapped out precisely, so the appearance of randomness arises. It should be noted that there is a clear distinction between 'random' outcomes and 'equally probable' outcomes. Plotting the outcomes of drawing different-coloured beads from a pot demonstrates random variation, but if the number of beads of each colour in the pot differs, there will be a pattern that emerges which clearly reflects the underlying population distribution given enough trials.

Students introduced to ideas of randomness by focusing on the outcome-based approach of theoretical probability and sample spaces may develop a deterministic view of probability, with a tendency to assume that a small sample, say six rolls of a regular die, should result in each number occurring exactly once as predicted by the sample space. Probability-based experiments can help students to build up a more complete picture of the mathematics of probability (see *Teaching Probability* by Jenny Gage and David Spiegelhalter, in this series), while developing a robust understanding of the key features of how randomness affects variation and how this can be accounted for to make meaningful inferences.

'Growing samples' activities are an excellent way of developing student understanding of how sample size has an impact on variation (Ben-Zvi 2006) and have been a mainstay of research on statistical reasoning in recent years. Students are asked to take a small sample from a population, say five data points, and draw a dotplot. The plots of all students in the class can be compared to demonstrate the wide range of shapes and average values that occur. Students can then be paired up and pool their data to give a sample size of 10; their new graphs will still be very different from those of other pairs, but some patterns will perhaps begin to appear – maybe a small range of values that seem to be covered in most samples, or a couple of extreme values that only occur once. This process can be repeated by combining the results of different pairs until a recognisable distribution becomes apparent in all the graphs. For the adventurous teacher, drawing students' attention to the fact that there may be some values duplicated in their shared data, and discussing whether this is significant, provides an opportunity to introduce the concept of bootstrapping, a modern inferential process of multisampling with replacement from a single sample to infer parameters of a larger population. This approach has become increasingly popular with the rise of computing power.

Figure 5.2 is a conceptual map of how the growing samples process works, highlighting the four main areas in which the nature of inference and the statistical literacy developed changes as the samples increase in size.

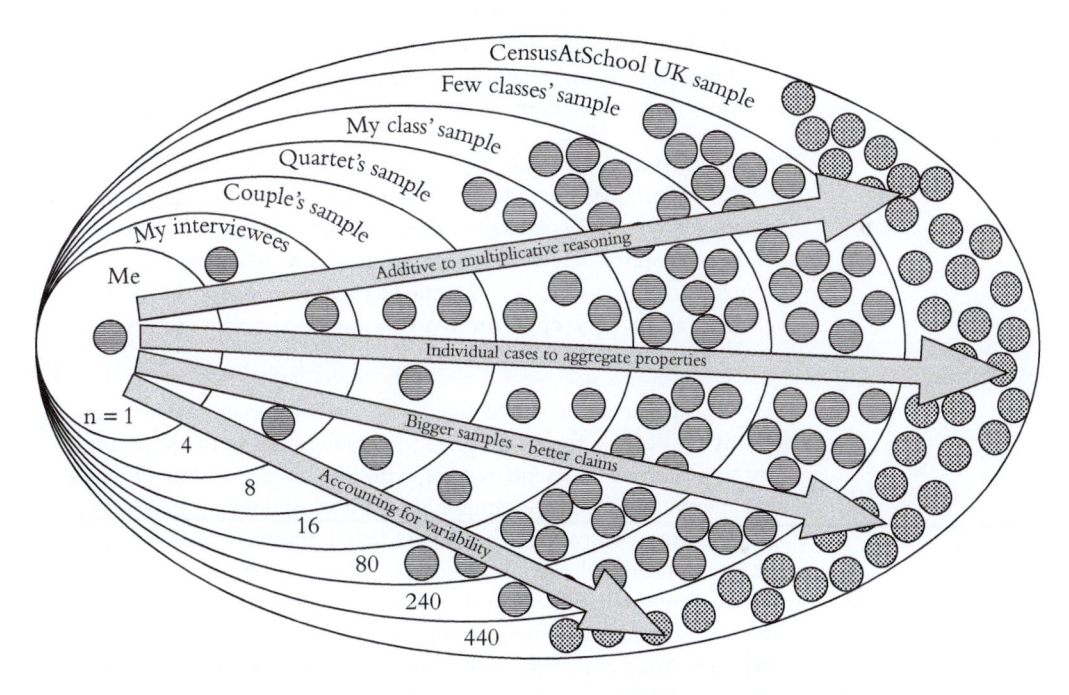

Figure 5.2 A growing samples sequence and the change in students' reasoning about informal inference
[Reproduced with permission from 'Scaffolding students' informal inference and argumentation', Ben-Zvi, D. (2006), in A. Rossman and B. Chance (Eds) *Proceedings of the Seventh International Conference on Teaching Statistics*. Voorburg, The Netherlands: International Statistical Institute.]

The 'growing samples' approach can help teachers neatly sidestep the time-consuming nature of data collection. Rather than asking students or groups of students to collect lots of data, each individual student can be tasked with collecting one piece of data, or contributing to one row of a multivariate data set. The classroom time freed up can then be spent on analysis and exploration of the data as the samples grow, enabling students to repeatedly make calculations, draw informal inferences and interpret the data. The key to this approach is to ask at each stage 'What do you think the data is telling you now?' and 'How have you changed your opinion based on the larger sample?'

5.3 Sampling bias

Bias in samples has an effect on the variation within a sample. Imagine a very crude example in which a literacy survey is sent out to households along with their subscription to 'You and your grammar' magazine (the no. 1 magazine for internet grammar pedants!). The likelihood is that the distribution of respondents will be overwhelmingly concentrated on the higher end of the literacy scale, with a relatively low standard deviation and little representation from the tail of people with weak literacy skills. Not only is the average literacy score likely to be higher than that of the general population, the variability of the sample will also have little in common with the variability of literacy rates across the population.

While many students are able to confidently describe sources of bias, this ability often does not extend to sound reasoning about sampling. Experiments with students suggest that they often prefer biased sampling methods such as self-selection to make things 'fairer' (Konold 2007). In this case 'fair' is used to describe the fact that respondents in the population all have the opportunity to choose to participate so no one's feelings will get hurt.

There also appears to be a tendency to over-rely on stratified sampling methods that seek to fix the proportions of subgroups in the population to try to guarantee representativeness, with little understanding of when this is appropriate. Stratified samples are a useful tool when certain conditions are met, crudely when the sample is relatively small, when there are proportionally small subgroups in the population and, crucially, when the proportions of each subgroup in the main population are known. If a student is taking a sample of 100 people, stratifying by gender in the knowledge of a 60:40 split, there is little need to stratify as random variation will likely provide a sample that broadly reflects the proportions of the population. If, however, the sample is of 20 people from a large population in which favourite music style is known to be an important factor for the object of study, it is worth stratifying to ensure that the small proportion who most enjoy Mongolian throat singing are not overlooked.

5.4 Target-error and population view of samples

Teaching about samples and sampling in lessons usually focuses on drawing a representative, random sample from a larger population, and it is this idea that most students associate with sampling. In further study, however, this conception of sampling does not encompass all aspects of the process or allow students to assess the appropriateness of different inferential techniques (Harradine, Batanero & Rossman 2011).

This population view of sampling has some further subtleties. Sometimes the sample will be drawn from a defined population, such as surveying a sample of students in a school. Other times, the sample comes from an infinite population, the most obvious instance of this being in probability experiments – an experiment in which a coin is flipped five times and the number of heads recorded can be repeated to generate a sample that represents the probability distribution, but there is no finite population that this sample is drawn from.

The target-error view of sampling arises from variation due to measurement, discussed previously. In this case, there is again no finite population, but there is an exact value which a process seeks, but fails, to replicate. Quality control processes often consider target-error; for example, a production line that aims to cut chocolate bars 7 cm long will produce an endless supply of tasty snacks that all measure close to the intended length. Analysing this distribution of lengths can give meaningful information about the intended length, if it is unknown, and the performance of the machinery over time. The samples generated for the Chapter 2 activity *Finding fault* are examples of the target-error view of sampling.

5.5 Onwards to inference

Understanding sampling and variation is a fundamental aspect of statistical literacy, and is essential both for students who do not continue to work with statistics beyond secondary education, and for those who go on to further study. For the former, a basic understanding of the ever-presence of variability and the outcome of biased sampling provides the tools with which to question authority. Asking 'Where did your data come from and how much does it vary?' is an excellent starting point from which to raise difficult questions. For the latter, formal statistical inference relies on a sound understanding of the effects and implications of variation and sampling, and will be discussed further in Chapter 7.

References and further sources

1. Moore, D. (1998). Statistics among the liberal arts. *Journal of the American Statistical Association*, 93, 1253–1259.

2. Reading, C. & Reid, J. (2004). Consideration of variation: A model for curriculum development. *Curricular Development in Statistics Education* (International Association for Statistical Education), 36–53.

3. National Research Council (1990). *On the Shoulders of Giants: New Approaches to Numeracy* (Washington DC: The National Academies Press).

4. Torok, R. & Watson, J. (2000). Development of the concept of statistical variation: An exploratory study. *Mathematics Education Research Journal*, 12(2), 147–169.

5. Gage, J. & Spiegelhalter, D. (2016). *Teaching Probability* (Cambridge University Press).

6. Ben-Zvi, D. (2006). Scaffolding students' informal inference and argumentation. *Proceedings of the Seventh International Conference on Teaching Statistics* (Voorburg, The Netherlands: International Statistical Institute), <https://iase-web.org/documents/papers/icots7/2D1_BENZ.pdf> accessed 7th February 2018.

7. Konold, C. (2007). *A Research Companion to Principles and Standards for School Mathematics* (National Council of Teachers of Mathematics).

8. Harradine, A., Batanero, C. & Rossman, A. (2011). Students and teachers' knowledge of sampling and inference. *Teaching Statistics in School Mathematics-Challenges for Teaching and Teacher* (Springer), 235–246.

Chapter 6

Signals and noise

6.1 Introduction

The two parallel threads in this chapter are: given a set of data, how do we see the 'signal' that the data conveys within the 'noise' of sampling, natural variation and so on; and how do we develop in students the ability to see distributions in an aggregate sense, so as to be able to draw out of a set of data not individual details or parameters, but an overall sense of the information within that set?

Distribution is one of the key concepts in statistics. Given a set of data, how do we see it as some sort of aggregate rather than simply a list, or table, of values? This process of learning to think in terms of aggregates is a known difficulty in the teaching of statistics (see, for example, Bakker 2004). Given the importance of graphical representations and summary statistics when working with large data sets and/or carrying out exploratory data analysis, this conceptual transition from case value to aggregate is worth taking some time to think about.

It is often the case that students know how to calculate measures such as means and medians, and know how to draw various types of graphs, but cannot interpret them or correctly choose when to use them. In fact, quoting single summary statistics or parameters, in frames which heighten the apparent largeness or smallness of the value, is a standard trick when using data to persuade rather than inform (for instance, the choice of absolute or proportional values in the media). We should strive to avoid this and to present data in a neutral fashion. One of the best ways of doing so is to represent, somehow, that aggregate sense, which is why data analysis is mostly about describing, or attempting to predict, aggregate features such as central tendency and spread. In this chapter we will consider how we can approach central tendency and spread less as summaries and more as descriptions of the signal present in the noise of the whole data set. This is often described as searching for signals in noisy processes (Konold and Pollatsek 2002)

6.2 Graphical representation and distribution

Ultimately, graphical representations are ways of visualising the structure or distribution of the data. Measures of central tendency or spread only have meaning in conjunction with each other, and in the context of the overall distribution. For example, if I tell you that the mean weight of an apple from my tree is 200 g, it does nothing to tell you how likely it is that I get an apple weighing more than 500 g. Similarly, knowing that the number of miles of driving I get from a full tank of petrol varies in a range of 100 miles 90% of the time tells you a lot about the variability, but very little about how many tanks of petrol it will take to drive the length of Route 66. We won't discuss skewness explicitly here, other than to think in a loose way about whether or not a distribution is symmetrical, and whether or not the tails are similar in shape.

The key thing is to make sure that statistical techniques and concepts are not taught in isolation, but are given a place in a coherent whole. Looking at distribution and comparing data sets via distribution gives a useful framework in which to learn the other techniques that lead more towards thinking in whole set or aggregate terms than algorithmic production of summary values (Cobb 1999).

6.3 Measures of central tendency

'Average', used colloquially, can refer to any one of several measures, such as the arithmetic mean, median or even mid-range. While each has its own uses and background, the key overarching idea is that the centre can only be meaningful as the centre of *something*, i.e. the distribution. It is not possible to decide which of the possible measures of centre might be appropriate without an understanding of the distribution of the data, and it is this sense of the average *within* the distribution that is the key to drawing students out of (or not allowing them to develop) the idea of these measures as simply the outputs of algorithms (Uccellini 1996).

> ### The mid-range
>
> The mid-range is not as well known as the mean and median and we will not discuss it much here. Because it is simply the mean of the largest and smallest values, this is the measure of central tendency that many students come up with independently if left to develop it for themselves. Of course the use of mid-range is entirely defensible in cases of symmetric and/or uniform distributions, but the simplicity of the measure, its sensitivity to outliers and skew, and its breakdown point of 0 (the proportion of arbitrarily large incorrect observations it can deal with without giving an incorrect result) do mean that it has to be used much more carefully than its simplicity suggests. However, it is a nice example of how much less use the central tendency of the data is without a sense of the distribution within which it sits; the mid-range either makes the assumption of a symmetric distribution or makes no allowance for the shape of the distribution, but there is no middle ground.

One way to go about this, following Uccellini, is to use manipulatives to develop the idea of the mean as representing an equal distribution of data. The use of cubes allows for division and distribution even when the mean is not a natural number. Begin with a distribution with a natural number mean; follow this with a mean that can be constructed by division of cubes, and then finally one that cannot be constructed concretely. The idea is to give students the blocks split unevenly into groups (say 2, 8, 4, 6, 3 and 7) and then ask them to balance the blocks between the groups. They will find that the blocks do balance out nicely, with 5 in each group. The number of blocks in each group is the mean and should be labelled as such so that the idea of it being the result of sharing out, or of equal distribution, is easily seen. This idea can be developed further by giving students groups which will not balance out with a whole number of blocks in each group. The discussion of how a fractional number of blocks can be representative of the number of blocks in each group, even if it is not physically achievable, is a key step in the development of the concept of mean.

Another way to think about the mean is as the balance point of a set of numbers, following an analogy with placing weights on a balance beam representing a section of a number line – or more correctly, given a weightless beam with equal-sized weights on it, where do you place the fulcrum so that the beam will balance? For example, given the numbers 4 and 6, it is easy to see that the balance point between the values is 5 since it is at an equal distance from each of the values (see Figure 6.1). However, locating the balance point gets harder with more values, and it is far easier to see that a given balance point is correct than it is to find that balance point intuitively. Given the values 3, 3, 4 and 6, where is the balance point? If you are told that the balance point is 4 then it is easy to see that two values are 1 unit less than 4 and one value is 2 units greater than 4, so the sum of negative and positive distances from the mean is zero as required. Thinking in terms of distances, you have to let the balance point be M and then solve the equation $2 \times (3 - M) + (4 - M) + (6 - M) = 0$, which is fine if you know the algebra, but quickly becomes unwieldy.

However, as a way to *verify* the position of the mean, and to understand the concept of the mean as a representation of the central point of the data, this balance point idea is very powerful. If it can be reinforced by physically working with weights and beams, then this is likely to be even better, if only because it allows for finding the mean experimentally and then generalising.

It is also instructive to allow students to develop general notions of measures of central tendency by estimation of large numbers. Coming up with reasonable ways of giving a representative value, then multiplying up, also allows for consideration of variation. Without this consideration

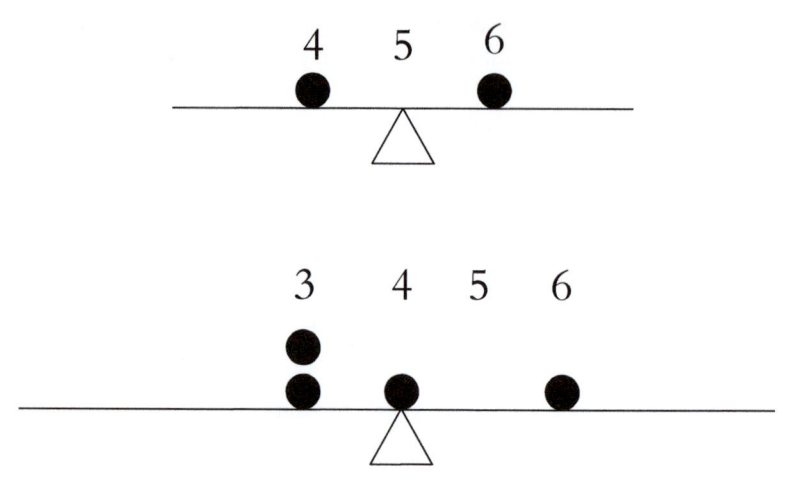

Figure 6.1 Representing the mean as a balance point

of variation, any estimate, and therefore any measure of average, is not a representation of the population. A good example activity, following a story of Thucydides as reported in Bakker (2003), is to count the number of bricks in a wall to estimate its height. This can be done in a school context anywhere with a large expanse of visible bricks, such as a sports hall or the outside of a classroom building. The problem is to estimate the height of the wall or building by counting the number of rows of bricks, then multiplying by the height of a typical brick. The average here is the typical brick height; the nice thing about this activity is that you start with the mean, or an estimate of it, and then multiply up to find the sum. This is the opposite direction to the usual algorithm for *finding* the mean, and it emphasises the role of the mean as being a representative value.

A group of students will get different values for the height and even for the number of rows of bricks. In order to get a group answer, they must extract from the noise of counting and measurement errors the signal of the 'true' number, on the assumption that the mean value will be pretty close. This action of repeated measurement, on either a group or an individual basis, is also instructive; there must *be* a 'true' value, but none of our values *is* necessarily that true value.

The context of repeated measurement, such as the brick counting activity outlined above, is also a possible route to developing the concept of median. When considering the errors in measurements, it is useful to compare them against the mean after the fact so as to demonstrate that they sum to zero, if only to draw on the idea of the balance point. By contrast, the median value, while most commonly thought about as the middle value of the data presented in an ordered list, can also be thought of as the value that minimises the sum of absolute errors, i.e. we are picking a value to represent the data by choosing that value which minimises the total absolute difference between the measured values and our chosen value. This is a reasonably accessible idea for students to consider, particularly as a follow-up to work around looking at the mean as a balance point, and may even lead to one of those 'wow' moments as students suddenly see a bigger mathematical truth to a procedure they have known since primary school.

It is worth noting that it is not very simple to show why this produces the same output as the median algorithm. However, this fact is easy to 'discover' using technology, particularly in a spreadsheet with a slider. Enter a set of data in a column in the spreadsheet and define a variable with a slider. You then find the sum of the absolute differences between each data point and the variable. By experimentation it is very easy to find the position of the slider which results in the smallest sum and verify that it is the median.

Adding a slider in Excel

Adding a slider, a button or some other form of control is done from the Developer tab in Excel. Just click on 'Insert' and 'Scroll bar', then onto the sheet where you want the slider. You will notice that the 'Design Mode' button in the tool bar toggles on. While this is on you can edit the slider; when it is off the slider will be usable. Right-click on the scroll bar that appears, then choose 'Properties', and you will see an entry for 'Linked Cell' in which you need to type the cell reference for where you want the value to appear. Excel will display in this cell whatever value the slider is currently taking, which you can use as a display, and also as a reference to take that value and use it elsewhere. You will also spot in the 'Properties' box places to change the increments so that you can define only integer values, numbers up to 2 decimal places or whatever is appropriate. Once you've finished playing, you will need to untoggle the 'Design Mode' button on the tool bar to check that the slider works; if it does, then when you move the slider up and down the value in the linked cell will change.

This is difficult to describe in words, but there are some excellent tutorial videos available online.

Another approach to the development of the median, within the distribution, is to get students to estimate the median from a dotplot or histogram. The problem here is to find the place in the diagram for which half of the dots are on one side and half on the other or, in the case of the histogram, half of the area is on each side. Again, this allows for development of the concept of median as representative without, or even prior to, introducing the standard algorithm for finding it.

Of course, the key fact about the median that makes it so useful in exploratory data analysis, namely its robustness to outliers, must not be forgotten. A lovely way to demonstrate this using technology, if you have already entered the brick counting and estimating data from earlier into a spreadsheet, is to display the mean and median in adjacent cells and then remove individual items of data at random, just to get a feel for the sort of variability seen in the two measures. You should see the median change in one of only two ways (what are they and why?), but the mean will vary by larger or smaller amounts depending on the distance the removed point is away from the mean. In particular, looking at data sets with noise, skew and outliers will tend to support the use of the median in the mind of students more than if your data sets are always neat and symmetrical.

6.4 Measures of spread

As previously noted, knowledge about measures of central tendency will only lead to understanding about data as an aggregate, rather than a series of individual cases, when thought about in terms of distribution, i.e. with some understanding about the nature of the variation around that value. Measures of spread are essentially measures of variation in the variable for which we have data.

Building on the example of the five-number summary earlier (see Chapter 3), an obvious measure to introduce early is the interquartile range (IQR). One of the great advantages of the IQR is its robustness or resistance to outliers. This is particularly true when working with experimental data generated by students, where samples often have to be small and time-efficient enough for students to generate their own data visualisations by hand. It is also very easy to sketch onto representations such as dotplots and histograms by area estimation, so that there is a rough-and-ready, easy-to-generate measure of spread to work with. Another reason for using and discussing IQR is its use in defining outliers, often identified as observations that lie more than 1.5 IQRs from the quartiles.

Total range is, of course, an option and it will often be suggested by students if they are allowed to generate their own measures of spread. The issue with the total range is the lack of robustness and the tendency to overemphasise the importance of the tails in normal-shaped distributions; this is a particular issue if we want the measure to help with identifying a data set as an aggregate, since some impression of the density of data, the presence of modes and so on is part of conveying a sense of the distribution rather than simply the interval in which values fall.

A mention also needs to be made of the interdecile range (i.e. the difference between the 10th and 90th percentiles), which has the effect of trimming (non-fat) tails and outliers while leaving behind the main part of the data set. This sort of trimming, even done informally, is an obvious next step once students have realised the effect of outliers on the total range, but before having the more sophisticated IQR to work with.

The standard deviation (or variance, or sample variance, or standard error …) has traditionally been left until later in the curriculum due to the complicated and time-consuming algorithm which students find difficult to reproduce error-free. Once again though, technology opens up avenues to explore standard deviation at a conceptual level without going into the details. Students who have explored the deviations from the mean can be told that standard deviation is simply a measure of the average distance of each value from the mean and should be able to instinctively grasp what a large value for standard deviation is in the context of specific data. Excel and other software with statistical functions will generate the standard deviation of a data set with no need for calculation; simply including this value along with other summary measures and graphs will help students develop a sense of standard deviation prior to meeting it in a more formal sense in further study.

Ultimately measures of spread used on their own have the same issue as measures of central tendency used on their own. In order to give an impression of the distribution of a dataset we need a range of descriptors and measures, and students should be encouraged to experiment with different ways of describing data to suit different contexts and purposes.

Variation due to measurement error has already been discussed, albeit briefly, in the example about brick counting and measuring. In this case there is a true value (or three values: the height of a brick; the number of rows of bricks; and the height of the wall), and the variation shows errors in the measurement of that true value. The true value will probably lie within the group of measurements, possibly discounting some outliers as large errors in measurement.

The other type of variation is that within a distribution; this shows the variation in cases of a particular thing. Here outliers *could* be measurement errors, but they can also simply represent interesting cases, different subpopulations and so on. The idea is to introduce the advanced concept of distribution early in the curriculum through the idea of shape; see for example Bakker (2004), Bakker & Gravemeijer (2004), Cobb (1999), or Russell & Corwin (1989). It relies to some extent on a guided reinvention pedagogy, by giving students data to work with and allowing them to do so in an intuitive way, relying on careful teacher guidance to encourage key developmental steps. The main preparation for an activity like this is for students to have a toolbox of possible graphical representations, or at least one or two which are appropriate for the data. As a useful fall-back position, dotplots have the advantage of being easily and quickly drawn by hand or by using free software like R.

Given a (small) set of data, students must choose a way of representing that set graphically. They look at their representation and decide whether or not it usefully represents the set. If it doesn't, then they are free to experiment with a different way of drawing it. Students can then reflect on what they have created and what signal it might allow them to infer from the noise of their data. You can easily combine this with a growing samples type of activity by giving each student a few data points from a data set which they can represent graphically before forming some sort of hypothesis or description of the data. Descriptions can (possibly should) be general rather than technical: 'It is sort of a triangle shape'; 'There are two bumps'; 'There are lots of values above the bump' and so on. Results can be compared with another student, sharing descriptions and critiquing them before combining their data so that they have twice as much. In coming up with a joint approach, students have to justify their individual choice of graph, come to agreement and synthesise their conclusions with those of their classmate. You can continue this process by combining pairs and so on, so long as the initial data set was large enough that no two students have overlapping sets.

Another approach is to collect a set of personal data, such as shoe size or height, from the class. This can simply be done by getting students to write the relevant number on a piece of paper and putting it in a box. Draw a series of small samples from the box, say of size 10, either giving each student their own sample or generating 4 or 5 samples and giving each student one of those. The same sort of process can now be followed, with students developing a representation, drawing conclusions about the signal given by their sample and then comparing with others to get a sense of the variation in the samples and the value (in terms of accuracy of signal) of having a large enough sample to get a proper idea of that variability.

Students will find it much easier to reason if they are given some prompts or an overarching problem to solve, rather than simply being left alone to perform unguided exploratory data analysis. For the activity *Shoe size* in Chapter 13, some simple prompts might be 'What size are most students' feet?', 'If I give a random student a pair of size 4 shoes, are they likely to fit?' or 'Can you predict the distribution of shoe size of students two years older than you?' These prompts can focus on case values, on distribution or on aggregation. It is important to give students a concrete focus, even if their work then goes off in different directions and begins to generalise and abstract.

All of this relies on an informal and free approach to the description of distribution. Students are very likely to work with case values to start with ('There are lots of people with size 6 feet') and to focus on outliers ('Wow, there is someone with *massive* feet!'), and so describe the noise rather than the signal. The more opportunity they have to work with reasoning and describing shapes, the more they move towards more use of adjectives and a more aggregate approach to distribution.

As a small postscript, it is worth noting that many data sets of the sort that you might measure with students or give to them as familiar examples exhibit right skew. This is usually to do with the fact that the values must be non-negative (or are physically bounded on the left somewhere), that small values are very likely, but that large values do sometimes occur. Examples include measured data like height and weight, but also income and number of children in a family. There are some useful left-skew distributions such as age at death and marks in an exam (assuming a reasonably good performance!), where the upper bound has a greater effect than the lower bound. For more interesting shapes you can create pretty good uniform distributions from scenarios such as dice rolling, and for bimodal distributions try book prices.

Generating summary statistics and standard graphical representations are important skills in most curricula, but without relating these to distributions and seeking to explain them in context of the aggregate data, it's like giving a comedian a series of punchlines with no premise to set them up. A median in isolation is about as illuminating as 'one says to the other "Can you smell fish?"' is funny, but providing the median as just one component of a broader picture allows a much richer picture to be established.

For those who need to know, 'Two parrots are sitting on a perch …'

References and further sources

1. Bakker, A. (2004). Reasoning about shape as a pattern in variability. *Statistics Education Research Journal*, 3(2), 64–83.

2. Konold, C. & Pollatsek, A. (2002) Data analysis as the search for signals in noisy processes. *Journal for Research in Mathematics Education*, 33(4), 259–289.

3. Cobb, P. (1999). Individual and collective mathematical development: The case of statistical data analysis. *Mathematical Thinking and Learning*, 1(1), 5–43.

4. Uccellini, J. C. (1996). Teaching the Mean Meaningfully. *Mathematics Teaching in the Middle School*, 2(3), 112–115.

5. Bakker, A. (2003). The early history of average values and implications for education. *Journal of Statistics Education*, 11(1), 17–26.

6. Bakker, A. & Gravemeijer, K. (2004). Learning to reason about distribution. *The Challenge of Developing Statistical Literacy, Reasoning and Thinking*. Eds D. Ben-Zvi and J. Garfield. (Dordrecht, the Netherlands: Kluwer Academic Press), 147–167.

7. Russell, S. & Corwin R. (1989). Statistics: The shape of data. *Used Numbers: Real Data in the Classroom. Grades 4-6* (Washington, DC: National Science Foundation).

Chapter 7

Informal inference

7.1 Introduction

The collection of data is not an end in itself, but a step in a process of modelling the world around us to help gain insight into systems or individuals. Analysis of large quantities of data by online retailers such as Amazon allows them to predict the kind of products you might be interested in buying based on a vast data set of customers and the combinations of items they have bought. These companies claim that recommending products provides a more streamlined customer experience, but it is not beyond technical possibility for an unscrupulous business to use the same information to adjust prices so that the unwary purchaser pays over the odds for the items that the model predicts they will most desire. A more sinister application of this kind of data collection and analysis is the targeting of adverts to individuals during political campaigns in order to influence voting patterns based on the specific triggers that models predict will have the most impact on an individual's decisions.

Many large companies' financial models are built on or enhanced by the ability to sell massive data sets harvested from their customers in order for other companies to mine the data for information and draw inferences about their target population. These practices give them tremendous power to target their products or messages.

So collecting data is not just about cataloguing what's happening now and in the past, but also a way of creating models to predict the future. Of course any model is imperfect, and a model based on a sample of the population is inevitably even more so. In order to account for the inbuilt errors of representing a population by a model based on a sample, sophisticated techniques in formal inference have been developed which students rarely encounter in their early studies.

The reason for this of course is that formal inference requires a deep understanding of the statistical process, as well as many technical mathematical skills. It is unreasonable to expect that students just beginning to study statistics should attempt to learn these formal techniques, but by exposing them to some of the key ideas, they are more likely to understand what is happening when they encounter formal techniques such as hypothesis tests in the future. It is these key ideas that we refer to as 'informal inference'. In Chapter 1 we discussed the omnipresence of variability and how it is fundamental to working with statistics. When attempting to make inferences from data, consideration of the implications of variability makes the difference between conclusions which say something meaningful about a system and conclusions which could charitably be described as bunkum!

In previous chapters, inference has not been discussed in much depth in order to concentrate on the topic covered, but in truth there is little point in learning anything about the nuts and bolts of statistics without considering how these skills can be used to make predictions. Inference is the process of moving beyond the data available either to test an existing hypothesis or to generate a new hypothesis. While most (but not all – more on that later) tests of a hypothesis are too technical to introduce to younger students, using data to generate new hypotheses is a first essential step on the rung to inferential reasoning.

> **Deduction or inference?**
>
> Pratt et al. (2008) described two 'games' which are undertaken when working with statistics. In Game 1, the dataset is all there is, whereas in Game 2 the dataset is a sample of a wider population. Game 1 is a purely descriptive game; any statistic is by definition also a parameter of the population, and definitive statements can be made. Game 2 is an inferential game; in this case, any statistic could be a reflection of the overall population, but there cannot be certainty any more.
>
> Asking students to describe whether they are playing Game 1 or Game 2 at any given time when undertaking statistical activities is a useful way of focusing students on whether they are describing or inferring, and signposting the necessity for a different approach. It also lays down some useful foundations for further study. Many people using statistics struggle to identify when it is appropriate to use Bessel's correction, dividing by $n-1$ rather than by n, when calculating standard deviation. It is rather simpler to say that whenever you are playing Game 2, Bessel's correction should be used.

7.2 Aspects of informal inference

Makar & Rubin (2009) claimed that:

'Focusing on investigating phenomena entails understanding the statistical investigation cycle as a process of making inferences. That is, it is not the data in front of us that is of greatest interest, but the more general characteristics and processes that created the data.'

Informal inference can be divided into three key aspects: probabilistic; generalisation; and from data (see Figure 7.1).

Statistical Inference

[probabilistic] [generalization] [from data]

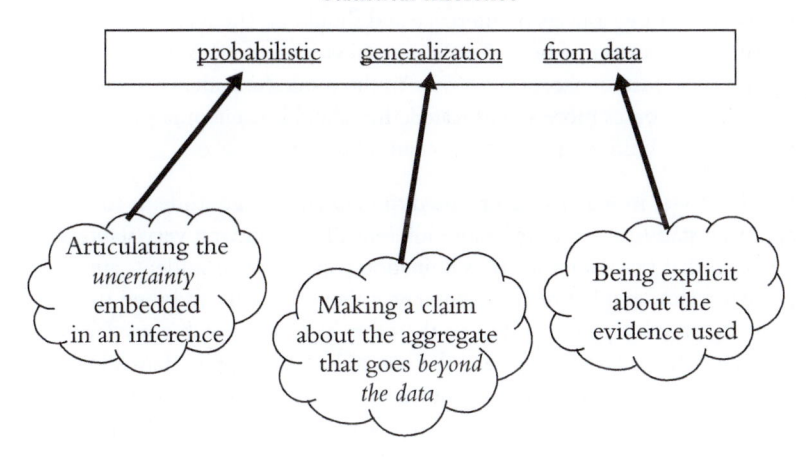

- Articulating the *uncertainty* embedded in an inference
- Making a claim about the aggregate that goes *beyond* the data
- Being explicit about the evidence used

Figure 7.1 A framework for thinking about statistical inference [From *A framework for thinking about informal statistical inference*, Makar & Rubin (2009)]

So an inference could be:

'By looking at the people that work in my corner of the office, I think that 2 in 7 people that work in this building are male, but I think it is unlikely that this is true.'

The sentence above has all the hallmarks of an inferential statement: it is based on data, the people who sit near me; it generalises to the population of all people who work in the same building as me; and it has a probabilistic statement – I have no confidence in the validity of my claim as I used a small sample that was likely to be very biased.

7.3 Structuring statistical conclusions

Many subjects use writing frames to help structure student responses, and this framework for statistical inference allows teachers to provide students with a checklist when making claims with data. This could easily be made to form part of a wall display so teachers can refer students to it each time they are working with data.

		Complete each Sentence
Data source	By looking at the data for …	
General statement	I think that …	
Confidence	I think my hypothesis is **likely/unlikely** because …	

Table 7.1 Writing frame for structuring statistical conclusions

Over time students will get used to thinking and making decisions in terms of the three aspects of inference and should be encouraged to articulate the reasoning behind their generalisations and also the reasoning behind their confidence, or otherwise, in the result. As students' inferential reasoning becomes more sophisticated, they should be encouraged to refer to specific techniques used in their chain of reasoning.

Small changes in practice can be used to help encourage students to approach statistical reasoning more inferentially. Consider a very standard question that often appears in examination papers where students are given a scatter graph, instructed to draw a line of best fit, and then use it to predict a value given a value on the other axis. It is worth considering carefully what is happening here; the line of best fit, happily drawn by students since primary school, is the creation of a mathematical model. This model represents the relationship between the two plotted variables, but it is far from perfect. Time should be spent discussing in class the strength of this model in relation to the data at hand: when is it valid and how good is the fit? For example, a strong positive correlation suggests the underlying variation is limited and therefore the predictions of the model are likely to be reasonably accurate, whereas a weak correlation may imply a much higher underlying variation and therefore the predictions of the model will be less reliable.

Using the framework above, students should be encouraged to make statements such as:

'Based on the model from my line of best fit, I think that a student who scores 72% in a maths test will score 63% in their English test, but most of the data does not lie close to the line so this is not a very reliable prediction.'

The statement above fits into the structure of the framework for statistical inference and forces the student to consider the strength of the model. Once students are used to thinking in these kinds of terms, they can be encouraged to quantify the variation further; for example, by considering two new parallel lines above and below the line of best fit, enclosing all the data, students could be asked to give a number for the variation. In the case of the example above, this could be:

'Based on the model from my line of best fit, I think that a student who scores 72% in a maths test will score 63% in their English test. All of the data lie within 12% of the line so the true value is likely to be between 51% and 75%. My prediction is not very reliable.'

In formal assessment this sort of subjective response about the quality of the model is not currently very common, but it is essential in class that students give consideration to such factors. This will help students engage in genuine reasoning, rather than simply reading a coordinate from an *x*-*y* plane. By thinking in terms of how the model relates to the underlying distribution, students will be better prepared when they meet more formal hypothesis tests later on in their study that use these kinds of ideas to relate a sample distribution to a theoretical model such as the normal or binomial distribution.

On occasion students may move from Game 1 to Game 2 during the course of an activity or an explanation, and setting the expectation that they always identify the game they are currently playing may help signpost when to use the framework for statistical inference. For example:

'Playing Game 1, for the population of students in our class, it appears from the boxplots that girls are generally taller than boys. Playing Game 2, I could say that for our school girls are generally taller than boys, but I wouldn't be very confident as my sample is biased because it is only looking at people in our class that are roughly the same age.'

By identifying which game is currently being played, there is a clear distinction in the above answer about when a description of the uncertainty needs to be applied to the conclusions.

The secondary benefit of following this kind of structured approach to statistical reasoning is that it encourages students to consider the context of the data along with the data itself. Researchers in education frequently observe students losing the connection between context and data as they undertake the process of data manipulation. In one experiment (Cobb 1999), students were tasked with investigating the lifespan of two brands of batteries: 'Always Ready' and 'Tough Cell'. The data were presented in a series of bar charts with green and pink bars representing the two brands. When discussing the charts, the students mostly referred to numbers and colours rather than battery brands, e.g. 'the green is usually about 110' rather than 'the Always Ready branded batteries usually last about 110 hours'. Students should always be encouraged to anchor their discussions in the context of the data rather than the more abstract representations of the data; this way the potential inferences are more easily identified.

7.4 Variability and inference

In Chapter 5 we discussed using a growing samples activity to help develop students' understanding of the power of sampling and the implications of sample size on variability. A second approach to this type of activity is to focus not only on how larger samples become more consistent, but also on how this consistency reveals the important truth that the shape of the sample distribution is more likely to reflect the population distribution as the sample size increases. While undoubtedly an essential thing for students to understand, there is a corollary to this fact that students are rarely forced to consider: even though this is generally true, probabilistic effects are still in play which muddy the picture. If every member of the population has an equally likely chance of being picked, it is no less likely that the smallest 100 data points will be chosen than it is to have a particular, mixed up, 'representative' selection of 100 specific data points spread evenly throughout the distribution.

For most people this is totally counter-intuitive, and you can test it very easily if you are prepared to ask your friends some annoying questions. Tell a friend that you have a bag containing coloured balls numbered from 1 to 100. You are going to draw five out of the bag at random, and then put them in order. Ask them which of the following results is more likely:

a   1, 2, 3, 4, 5

b   2, 17, 24, 46, 73

Unless your friends are statisticians, or were really, really paying attention at school, their answer will probably be b). We all have a tendency to fall back on the 'representativeness heuristic', a mental schema that, in this case, tells your unsuspecting friend that a jumbled string of numbers looks more like what you would expect to draw at random from the bag, and therefore is more likely to occur.

So what is really going on here? The two strings of numbers are equally likely – both probabilities are calculated by the same product, because the probability of each ball being drawn is 1 out of the remaining balls in the bag:

$\frac{1}{100} \times \frac{1}{99} \times \frac{1}{98} \times \frac{1}{97} \times \frac{1}{96} \approx 1$ in 9 million chance.

The key is that when we draw balls from the bag, we equate any jumbled up mix of numbers (for which there are millions of possible combinations, each specific example of which has the same probability) with sample b) and therefore assign a higher probability to this situation, but for the specific values in both sample a) and sample b) the probability is the same.

It is rare for students to be made to consider the implications of this, that a single random sample may not be representative of the population at all, but a freak sample in which all the small values in the population were chosen. From a basic understanding of probability it is clear that if every data point in the population is equally likely to be selected, every once in a while all the data points will be selected from an unrepresentative area of the population distribution (the hypothetical graph of all possible data points).

There are good, historical reasons why students have not previously looked at repeated sampling. It is not something that most people find intuitive, and with the limited tools available in previous decades, most people relied on equating samples to theoretical models of distributions via formal mathematical techniques. While students do eventually come across these ideas in higher study, it is often as part of a mechanical process that is ultimately boiled down to feeding data into a formula and turning a metaphorical handle before comparing the sizes of two numbers and making a standardised statement.

The alternative, to make a class of students each take an independent 'representative sample' and compare them, is unlikely to generate the highly unlikely unrepresentative samples. In most cases, a set of 30 samples taken from a larger population will all still appear to be comparable despite some variation. Even if the probability of a sample containing mostly data from the tail is only 1 in 100, those are not good odds when planning a lesson to explore this feature.

With modern software, using tools such as www.lock5stat.com/StatKey, real data sets that students have been working on can quickly be sampled thousands of times to create a sampling distribution, and students can be challenged to make comparative statements between, for example, the mean of their own sample and the mean of the sampling distribution. It should be noted of course that a great deal of similar sounding but profoundly different terminology can creep in, and teachers should be careful to avoid using technical language in these early experiences of inferential reasoning. Students should not need to learn the difference between a sampling distribution, a population distribution and a sample distribution in order to experience the implications of these things.

7.5 Visualisation software

Much of the recent research on informal inference has involved small-scale studies with younger students, using dynamic statistics visualisation software such as *Tinkerplots* (www.tinkerplots.com) and *Fathom* (fathom.concord.org). These are software packages specifically designed to allow interactive exploration of multivariate data sets by creating standard and non-standard visual representations of variables of interest. A wealth of research papers, many of which can be found on the Tinkerplots website, suggest that the availability of these kinds of software packages allows teachers to create an environment in which students can explore data and be encouraged to create new hypotheses for further study as they progress. One such example (Paparistodemou & Meletiou-Mavrotheris 2008) details students' exploration of data that had been collected from a survey of students in their school and input into the software. Students initially created a graphical representation of the answer to the question 'Do you play with scissors', and were able to read from the graph that most children, possibly through a combination of Pavlovian training, inherent common sense, and the iron discipline of a teacher terrified of classroom injuries, did not in fact mess around with craft tools.

At this point, the students reportedly began creating conjectures of their own, that perhaps the few students who were foolish enough to risk life, limb and minor injury by playing with scissors were more likely to be in a lower school year, as they had not been taught yet about the importance of safety in the classroom. By manipulating the variables on the axis, the visual display could be easily reconfigured from a simple bar graph of yes/no to a compound bar graph with three bars representing the different school grades, each bar showing the breakdown of safe and potentially dangerous students.

Several benefits have been observed in this approach to developing inferential skills. Students are often more engaged by forming their own questions and conclusions, and are able to explore data in ways that they find personally interesting. They get used to creating a conjecture and then thinking about how the data must be arranged in order to gain some insight. Finally, the software representations are adjusted dynamically in ways that are not always predictable, so students get used to interpreting diagrammatic representations by attempting to read them, rather than by fitting them against a standard mental template which may be quite restrictive.

The skill of the teacher in this sort of approach comes from selecting, or guiding the students towards creating, a multivariate data set rich enough to allow students to explore without hitting a brick wall of needing extra variables in the data to investigate further. Additionally, teachers must carefully monitor student responses and provide guidance to help students refine their questions, conjectures and conclusions, ensuring that these are based on evidence from the data rather than their own preconceived opinions and ideas.

Moving towards this approach can be challenging for both teachers and students, and it is reasonable for teachers to have a healthy scepticism of the kind of independent, project-based, open-ended tasks encouraged by many educational researchers. In a real-world classroom, teachers may question the ability of the students to conduct exploratory work unsupported and be concerned about the workload of marking project-based assignments. Additionally, access to computers and software costs can be a barrier.

To overcome this, teachers can use a guided exploration approach in which the students are responsible for making decisions, reading data from visualisations and arguing for the validity of their conjectures, while the teacher is responsible for displaying and manipulating the data as instructed by the students. In this way, the technical limitations are reduced and the students themselves can provide the ideas and the critique, explaining to their classmates what each visualisation shows and suggesting fresh avenues of exploration. Teachers can provide immediate feedback and step in when an insecure conjecture goes unchallenged by the remainder of the class.

Both *Tinkerplots* and *Fathom* host a number of data sets on their websites to be used with the software, along with links to external data sets in more open formats. These can be taken as a starting point for exploration, or used in conjunction with a wealth of hosted lesson activities that can be used directly or adapted to suit. Wild et al. (2011) suggest an approach to developing inferential skills that focuses on working graphically and ensuring that students' attention remains on the visual representation as much as possible rather than attending to calculations and other 'busy work' (see Figure 7.3 overleaf). It is important that in lessons and activities designed to develop inferential reasoning, students remain focused on the big picture, rather than getting distracted by activities like constructing graphs by hand or calculating statistics. As a result, data representations and summaries should be constructed using technological tools, or provided to students as pre-printed material.

Boxplots are an excellent introductory tool for students to work with and can be further enhanced by overlaying a dotplot of the original data, as boxplots can be quite abstract items (see Figure 7.2).

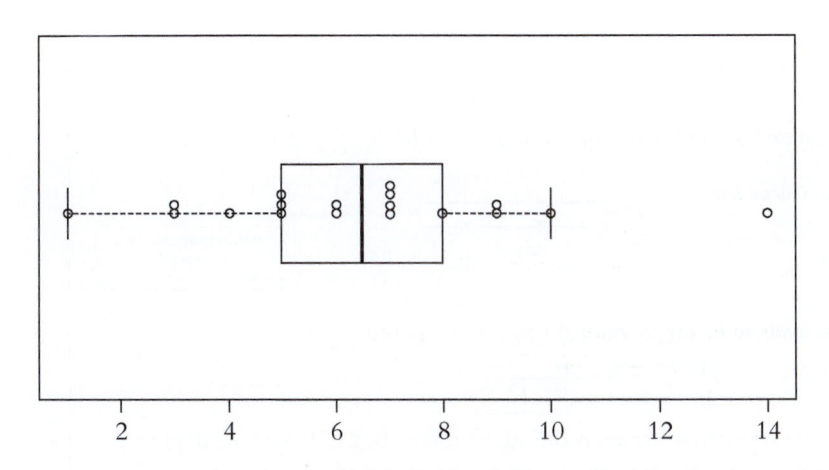

Figure 7.2 Dotplot overlaying a boxplot

An explanation of how to construct these kinds of graphs using R can be found in Appendix 1.

Guidelines on "how to make the call" by development level

At all levels:

A ⊢─□▯─┤
B ────⊢▯□⊢──

If there is no overlap of the boxes, or only a very small overlap make the call immediately that **B tends to be bigger than A** back in the populations

Apply the following when the boxes do overlap ...

Milestone 1 test: *the 3/4-1/2 rule*

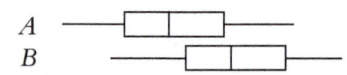

A ⊢─□▯──┤
B ────⊢▯□⊢──

If the median for one of the samples lies outside the box for the other sample
(e.g, *"more than half of the B group are above three quarters of the A group"*)
make the call that **B tends to be bigger than A** back in the populations

[Restrict to samples sizes of between 20 and 40 in each group]

Milestone 2 test: *distance between medians as proportion of "overall visible spread"*

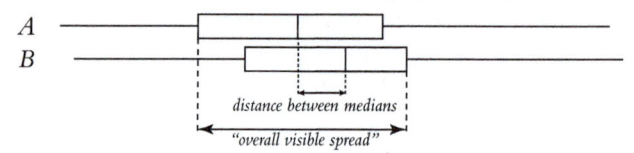

A ──⊢─□▯──┤──
B ──⊢──▯□─┤──
      distance between medians
    ◄── *"overall visible spread"* ──►

Make the call that **B tends to be bigger than A** back in the populations
if the distance between medians is greater than about ...

⊢──┼──┼──┤
1/3 of overall visible spread for **sample sizes of around 30**

⊢─┼─┼─┼─┤
1/5 of overall visible spread for **sample sizes of around 100**

[**Could also use** 1/10 of overall visible spread for sample sizes of around 1000]

Milestone 3 test: *based on informal confidence intervals for the population median*

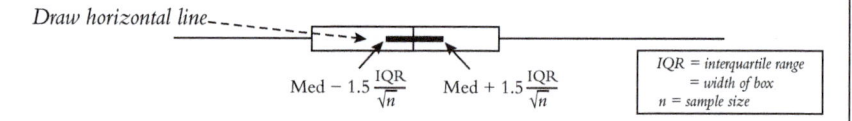

Draw horizontal line- - - →

Med − 1.5 $\frac{IQR}{\sqrt{n}}$ Med + 1.5 $\frac{IQR}{\sqrt{n}}$

IQR = interquartile range
    = width of box
n = sample size

Make the call that **B tends to be bigger than A** back in the populations

A ────⊢□┼□┤────
B ──────⊢□┼□┤──

if there is compete separation between the added intervals (i.e. do not overlap)

Milestone 4:   *on to formal inference*

Figure 7.3 'How to make the call'
[Reproduced with permission from C.J. Wild, M. Pfannkuch, M. Regan, N.J. Horton, Towards more accessible conceptions of statistical inference, *Journal of the Royal Statistical Society: Series A (Statistics in Society)*, 174(2), 247–295 (John Wiley and Sons)]

The approach recommends that students consider the obvious visual changes on each subsequent boxplot when taking repeated samples of data from a population. The intention is to help students develop an appreciation that when working in Game 2, a single boxplot is in fact an instantaneous snapshot where each sample generates a slightly different view of the data. Visual representations of this for use in class can be found at https://www.stat.auckland.ac.nz/~wild/WPRH/. Alternatively, asking students to each take samples of different sizes from the same population and consider the differences and similarities is a lower-tech approach. The key is that when students picture their boxplot, their mental schema when working in Game 2 has some fuzziness and uncertainty at each point in the five-number summary, and the size of this fuzziness is related to the size of the sample that was used to construct the plot.

Once students have a clear conception of the effect of sampling variation, a more concrete trajectory can be implemented to develop their inferential decision making. Wild et al. (2011) proposed that students' decision-making with regard to comparing boxplots and inferring whether the data values for population B were larger than for population A could be developed using their guidelines on 'how to make the call' (Figure 7.3).

Students begin by only declaring B to be larger if there is no overlap in the boxes at all (or only a very slight overlap). Once they are comfortable with this, they can begin to make more technical judgements based on the amount of overlap between the boxes, initially only when the median of one sample is outside the box of the other, before looking at differing proportions of overlap dependent on sample size as they become more adept. For students who may be expected to go on to study statistics at a higher level it may be appropriate to explore more formal calculations of confidence intervals, but this is beyond the scope of most current secondary schemes.

Clearly this kind of developmental approach requires the skills to be revisited and repeated over a period of time, and so opportunities should be sought to allow students to work regularly with real data to generate these analyses.

7.6 Making the most of probability

Although tentatively grouped together in a large number of curricula, the intersection of the Venn diagram which contains probability and statistics is often left unexplored until students encounter hypothesis testing during further study (see Figure 7.4). Even then the probabilistic foundations of the models used for such tests are often left unexplored. In lower

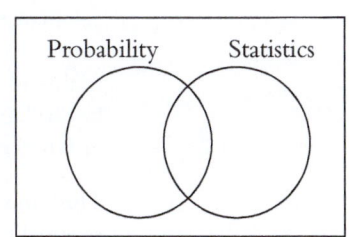

Figure 7.4 Venn diagram of probability and statistics

secondary school teaching, probability can be a useful tool for developing inferential reasoning skills situated in contexts for which a definite answer is available.

In *Local and global thinking in statistical inference,* Pratt et al. (2008) examined how students interacted with a software minitool called *ChanceMaker* that simulated a 'wonky dice' and how they developed understanding of the underlying parameters of the simulation. A statistical approach to probability experiments has potential advantages in terms of students' understanding of probability. A common pitfall when students are introduced to probability by focusing on theoretical combinatoric methods is that they can develop a somewhat deterministic view of random phenomena, expecting the experimental distribution to reflect precisely the theoretical for a small number of trials. A naïve conception of probability is common: asking students what they would expect to happen if a standard die were to be rolled six times will often produce the response 'there should be one of each number'. In this case students lack the awareness that the theoretical probability is only an indicator of the overall distribution for a large number of trials, with the proportion of each outcome approaching the theoretical value as the number of trials approaches infinity. While this is an important concept, it can be approached more generally by encouraging students to make inferences from probability experiments about the construction of the generative tool being used, be it dice, spinner, minitool or some other gizmo.

One possible approach is to use a hidden tool to generate results and ask students to recreate it, then test if their inferred tool produces similar results. Use a spinner or a Bernoulli pot (a bag containing coloured beads) and repeatedly trial to generate data. Encourage students to record the data and develop graphic representations from which they can make predictions about the proportions of the spinner or the coloured beads. You could structure this to ask students to make predictions after 5, then 10, then 15 trials, and so on, or you could allow students to make their predictions as soon as they have convinced themselves that they are likely to be correct and explore the implications of this.

Once students have made their predictions, without revealing the parameters of the generative tool instruct students to create their own version of the tool based on their prediction, and then run their own trials with this model. Once they have generated some data they should be challenged as to whether they think the data backs up their prediction, or whether they are no longer confident that they were correct.

This kind of classroom activity introduces many concepts that are essential to formal inference, laying the foundational ideas of probability distributions, statistical models, and comparing experimental data to model data, along with both the law of large numbers and the dangers of hasty generalisation from small numbers of trials. Depending on the students, it may be appropriate to explore some of these concepts in more detail and add some mathematical gloss to their informal reasoning.

7.7 To inference(ity?) and beyond

In this chapter, the key ideas underpinning formal statistical inference have been explored using informal approaches accessible to most students. While the conclusions that students come to may ultimately be slightly arm-wavy and imprecise, the conceptual foundations they are building will allow them to develop more formal skills either during further study or in a context relevant to the work or study they are doing. Arguably more importantly, a deeper understanding of the basics of inference will prepare students to be critical consumers of information throughout their whole lives.

References and further sources

1. Pratt, D., Johnston-Wilder, P., Ainley, J. & Mason, J. (2008). Local and global thinking in statistical inference. *Statistics Education Research Journal*, 7(2), 107–129.

2. Makar, K. & Rubin, A. (2009). A framework for thinking about informal statistical inference. *Statistics Education Research Journal*, 8(1), 82–105.

3. Cobb, P. (1999). Individual and Collective Mathematical Development: The Case of Statistical Data Analysis. *Mathematical Thinking and Learning*, 5–43.

4. Paparistodemou, E. & Meletiou-Mavrotheris, M. (2008). Developing young students' informal inference skills in data analysis. *Statistics Education Research Journal*, 7(2), 83–106.

5. Wild, C. J., Pfannkuch, M., Regan, M. & Horton, N. J. (2011). Towards more accessible conceptions of statistical inference. *Journal of the Royal Statistical Society: Series A (Statistics in Society)*, 174(2), 247–295. <https://doi.org/10.1111/j.1467-985X.2010.00678.x> accessed 6th November 2017.

Tinkerplots can be downloaded from www.tinkerplots.com

Fathom can be downloaded from https://fathom.concord.org

ChanceMaker can be downloaded from http://sites.google.com/view/davepratt

Chapter 8

Technology in the classroom

8.1 Introduction

Many of the advances in statistics and data science have developed in parallel with the rise of computing power. Where once the volume of data that could be processed was limited by the time it took for an individual or a team of people to calculate by hand, there is now no effective limit on the amount of data that can be processed. The hard limits that data scientists face are now based around the ability to store the vast quantities of information produced, and even this factor is not remaining static.

It is a shame, then, that while the discipline of statistics is moving on at an ever accelerating rate, the activities that take place in the classroom do not reflect this. While we would have loved in this book to write at length about the modern statistics that could be taught by relying on statistical analysis software, a certain amount of realism is necessary. In most schools, resource restrictions and access to training make it very difficult to radically change the approach to statistics. For policy makers, these same restrictions hinder the adoption of new, technology-dependent processes in the curriculum and assessment frameworks. It would be a foolish minister who demanded that all schools suddenly begin teaching their students technology-dependent statistics when many schools do not have the ICT infrastructure to deliver the lessons or the opportunity to train teaching staff in the use of new techniques, software and pedagogy required.

In this chapter, we will provide a brief overview of some practical approaches to incorporating statistical software tools into the teaching of current curriculum material, in order to help teachers become familiar with the opportunities such software affords, in preparation for a future in which it will one day be the norm in classrooms. In previous chapters we have mentioned *Tinkerplots* and *Fathom* as useful tools. Although we would recommend these as excellent exploratory learning environments, there is a cost associated with the licences, so many teachers reading this book will be unable to access them in a classroom environment. As a result we will not discuss these pieces of software in this chapter but gently suggest that they should not be forgotten by centres with the resources available to explore them further. Both pieces of software are intuitive to use and have excellent tutorial material available as well as restricted-functionality demos that the interested teacher could investigate further.

Why are computers so good at statistics? Because they 'R'

R is a free, open-source statistical analysis package widely used in the 'real world'. There are some excellent reasons to consider introducing this software into classrooms.

1. If you can use R to create visual representations, many lesson planning tasks that are time-consuming using the functions in Excel would be considerably simplified. Things which would need preparation in advance can be done on the fly in lessons, which means you can use more 'real data' collected by students.

2. Working with data in R is very similar to coding using functions. In the past, this would have been a huge barrier to students accessing the software, but with more and more countries raising the expectations of coding in the curriculum, soon students will be used to working with the kind of interface R provides.

3. R, alongside other packages with similar methods of interaction, is widely used by real-world data analysts. For any students who will work with data in future careers or education, some experience of R is likely to provide them with a very useful foundation.

4. There are some excellent graphical environments built on R, such as 'shiny', that can provide excellent demonstrations of statistical principles in class and a rich source of project-based material for students familiar with the language of R.

In this chapter we will discuss how to install R and conduct a random sampling demonstration for students. We will also contrast this with how the same activity could be run using an Excel spreadsheet. In addition, we will introduce *Gnumeric*, a free alternative to Excel with greater emphasis on statistical functions. Appendix 1 contains a selection of useful R commands for creating graphical representations and statistics from data.

8.2 Getting started with R

R can be downloaded from www.r-project.org. Once you have downloaded the file for your operating system, the executable installer can be run. For teachers installing on a school network, you may need permission or assistance from your IT technicians.

Once installed, run R from the icon on your desktop (see Figure 8.1).

While data can be imported manually into R, in most cases, you will want to import data held in a spreadsheet. In order to do this you must

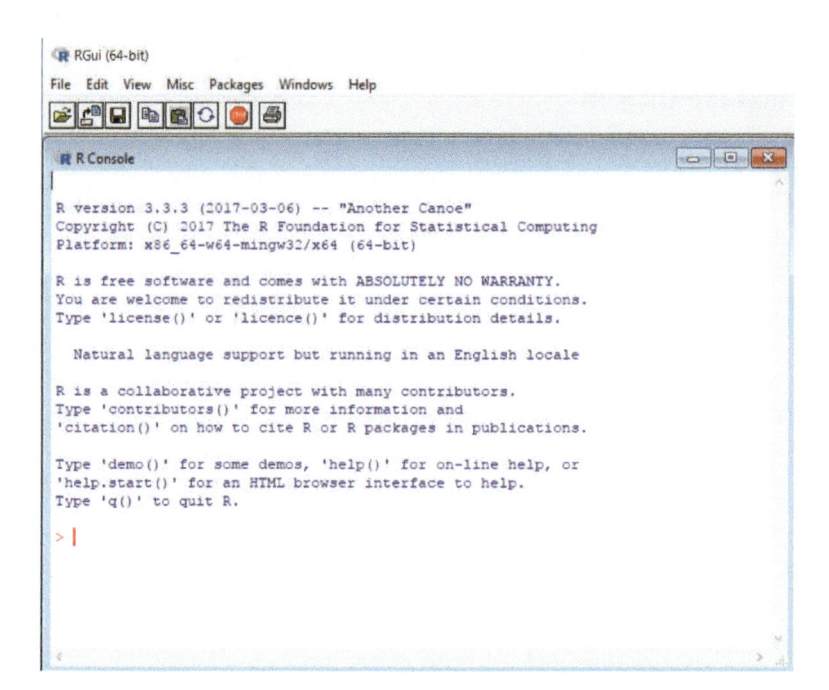

Figure 8.1 The R user interface

save any data sources into R's working directory. The working directory can be found by typing the command **getwd()**, which will tell you the default location. If you are happy with this location, simply make sure you save any data files to this folder on your computer. If, however, you would prefer to change the location, you must use the command **setwd()**:

setwd("c:/Users/myRData")

In the example above, the working directory is set to a folder called myRData, a sub-folder of users on the c: drive.

It is worth noting a few important syntax points here. Most actions in R occur when you call a function; in this case, the functions we have used are **getwd()** and **setwd()**. All functions are followed by parentheses into which additional parameters called 'arguments' are typed, such as the location of the new working directory for **setwd()**. In most cases, if the parentheses are left empty a set of default arguments are used. This can appear a little complicated initially, but after a short time you will begin to recognise similar arguments in many functions; often these are interchangeable. One other important thing to remember is that R is case sensitive, so make sure you keep track of upper and lower case letters.

Once you are ready to import some data, enter it into a spreadsheet and save as a .csv file, a simple, universal format that can be selected from a drop-down menu in most spreadsheet software.

To import the data into R use the following command:

```
yourData<-read.csv("yourFileName.csv",
header=TRUE)
```

Taking this step by step, **yourData** is the name we are giving to the object that your imported data will be stored in; this can be named whatever you like. Whenever you manipulate your data you will use this name. If you are importing several data sets, make up a descriptive name for each one.

read.csv() is the function that pulls the data out of the csv file and drags it into R. The arguments in the parentheses are firstly the file name in quotation marks, followed by a comma which is always used to separate the various arguments, and then **header=TRUE** which tells R that the first row of your data is the name/description of the column; if you do not have column headers, change this to **header=FALSE**.

8.3 An example activity using R

A useful demonstration is to show students how a sample taken from a larger population does not always represent the total population. One way to do this is to take repeated samples from a bivariate data set and plot the data in each sample as a scatter graph. With each new sample, the pattern in the data can look remarkably different.

To set up this demonstration in R, import your chosen bivariate data using the method detailed above. If you type the name of your data set, it should appear in the window as a list. For this example we are using a classic data set consisting of durations of eruptions of the geyser 'Old Faithful' in Yellowstone national park, which is one of several data sets pre-loaded in R and can be found by simply typing **faithful** into the R console.

In the rest of the example we will assume that the data has been named 'yourData'.

To draw a scatter graph of the whole data set, simply type

```
plot(yourData)
```

and a scatter graph should appear in a new window (see Figure 8.2). Plot is a very clever function that takes an educated guess based on the data type as to what kind of graph to draw.

This scatter graph plots the waiting time between eruptions against the duration of the following eruption.

We want to look at repeated samples, rather than the whole data set, so we need a new function. Fortunately

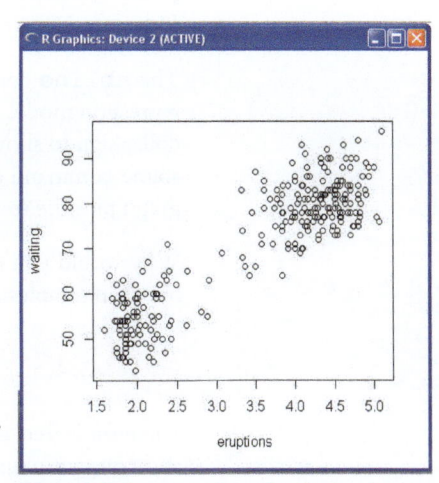

Figure 8.2 Scatter graph produced using the 'plot' function in R

R has a straightforward approach to naming useful functions: we want to take a sample, so we use the **sample()** function as an argument in our plot:

`plot(yourData[sample(272,10),])`

So what is going on now? Looking inside the parentheses, we have started with the old **plot(yourData)**, but instead of having just **yourData** in the parentheses we now have **yourData[sample(272,10),]**.

The function will now draw the plot from a sample of yourData, but what do the numbers do? In this example, the population size is 272, and I want to take a sample of 10 from it. One other thing to notice is the extra comma before the square bracket at the end; this must be left in or the function won't work.

Pressing the up arrow duplicates the previous line, so to update the sample and show multiple graphs one after the other simply hit the up arrow followed by the enter key.

Figure 8.3 shows four graphs produced, very quickly, with this repeated sampling technique.

These four *mostly* show the same two subpopulations as occur in the population set, but not consistently. Note that R will automatically pick best-fit axes *for each sample,* so the scales on the axes are not necessarily the same if you produce a series of samples in this way.

You can easily add a trend line, effectively creating a linear regression model and then plotting it, using the command

`plot(yourData[sample(272,10),])+abline(lm(yourData$waiting ~ yourData$eruptions))`

The **abline** command will draw a line for you, the **lm** generates a linear regression model for the two variables defined in the brackets using the dollar sign to show which data set to look them up in. Appending this abline command using a plus sign allows you to draw the whole scatter graph plus the trend line in one go so that it is easy to refresh the sample.

Why would you do this? Because the trend line will vary massively between samples and is a reminder of the variability within small samples.

8.4 Doing it again with Gnumeric

Gnumeric is free, open source, spreadsheet software available since 2001 and constantly updated. It is developed primarily for Linux-based computers and features regular updates on this platform. For more

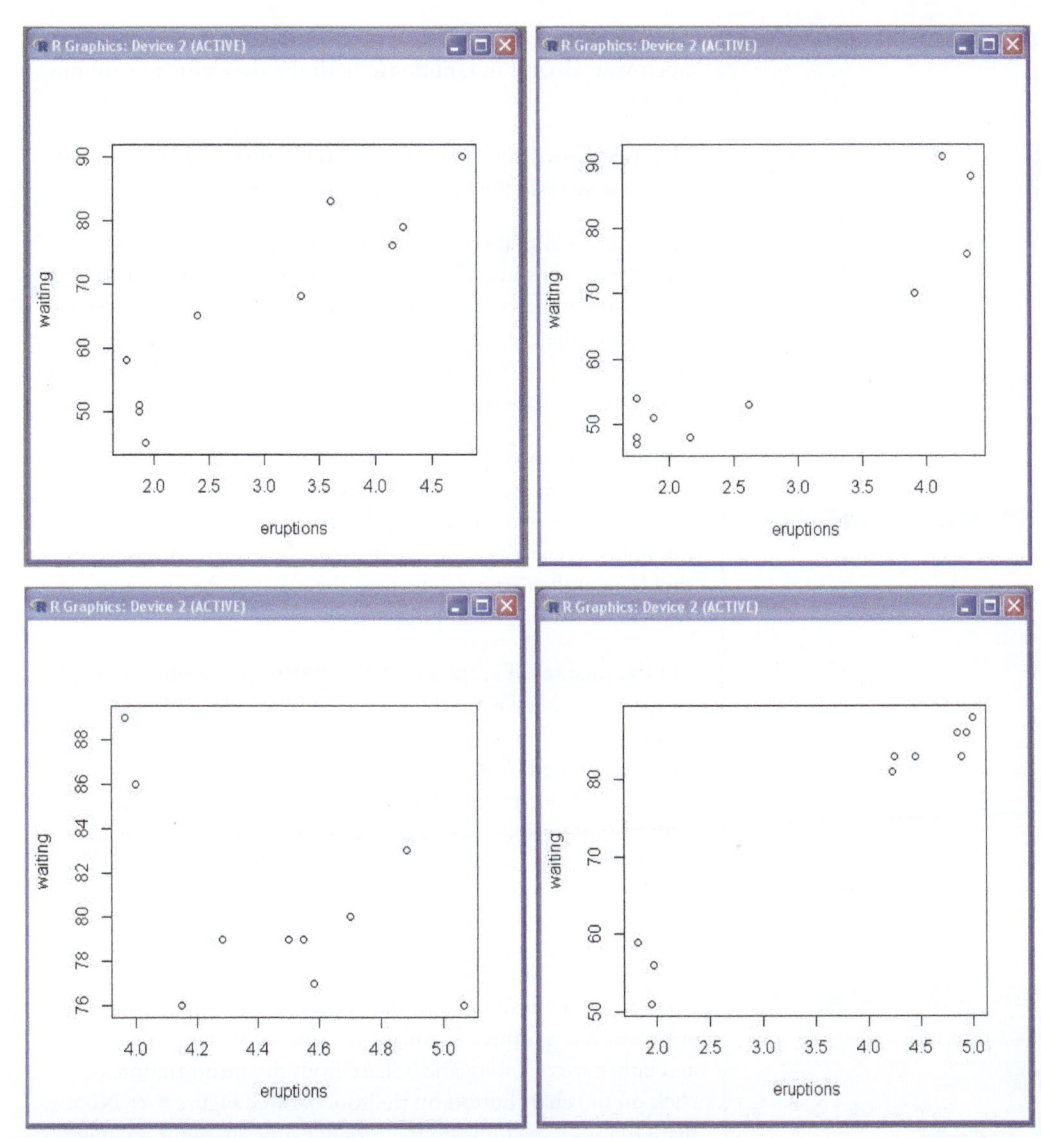

Figure 8.3 Scatter graphs of four samples of size 10 generated in R

common Windows-based computers, the most recent version available at the time of writing is v1.12.17 from 2014. This version can be downloaded from www.gnumeric.org and numerous other trustworthy software download sites.

Once Gnumeric is downloaded and installed, you will see a window that should be familiar to anyone who has used spreadsheet software before. The advantage of using Gnumeric is that it has additional statistical functions built in, such as the ability to sample from a dataset, that are unavailable in some commercial spreadsheets.

To reproduce the sampling activity in the previous section, open your dataset in Gnumeric with the data pairs in columns A and B.

Select 'Statistics' and 'Sampling' from the menu bar, which will open a new dialogue box (see Figure 8.4).

Select the cells that contain your data. A good trick which works in many spreadsheet packages is to select the first row of data and then press ctrl + shift + down arrow. This will select all the data in the column without having to scroll and drag. Make sure 'Grouped by columns' is selected, and if your columns have titles tick the 'Labels' box.

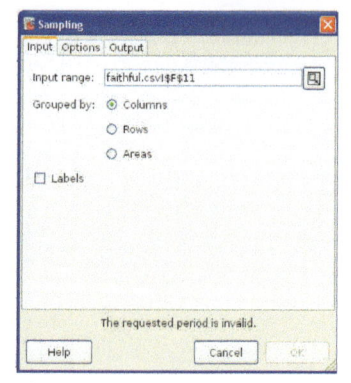

Figure 8.4 The Input tab of the Sampling dialogue box in Gnumeric

On the next tab you will set the options for your sample (see Figure 8.5). Set the sampling method to 'Random' and choose the desired sample size. Depending on how you are intending to use the data you may want to take more samples simultaneously; this can be set on this tab, but for our purposes we will take a single sample.

On the final tab (Figure 8.6), select where you want the sample to appear. To keep everything on one page, select 'Output range' and click in the top left cell where you want the data to be positioned. You can decide whether 'Enter into cells:' is set to 'Formulae' or 'Values'. As we are making a dynamic demonstration for a class, we will use 'Formulae' to allow the sample to be updated in real time later.

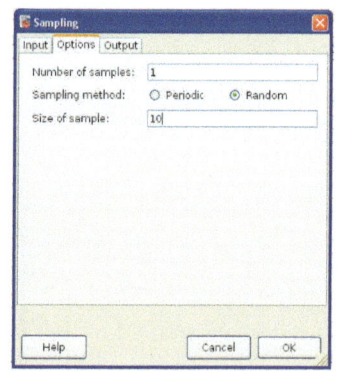

Figure 8.5 The Options tab of the Sampling dialogue box in Gnumeric

Once you are happy with your choices, simply click 'OK' and the sample will appear in your spreadsheet.

A graph can be inserted into the page in the same way as for any spreadsheet: simply highlight the cells containing the sample and either select 'Insert' and 'Chart' from the menu ribbon, or click on the chart button on the tool bar (see Figure 8.7). Note that Gnumeric will update the axes dynamically, but it will rarely truncate them.

To resample the data, press F9 to refresh the sample and watch the graph update.

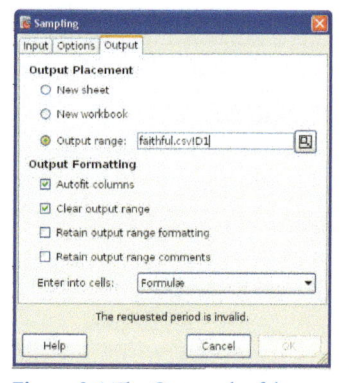

Figure 8.6 The Output tab of the Sampling dialogue box in Gnumeric

To add a trend line, double click on the graph to bring the graph settings window up (see Figure 8.8). Select 'PlotXY1' and click 'Add'.

From the drop-down menu that appears select 'Trend line in…' and 'Linear' for a straight line, then click 'Apply'. As you resample the data, the trend line will also update dynamically, making it

PART 1 A vision for statistics in schools

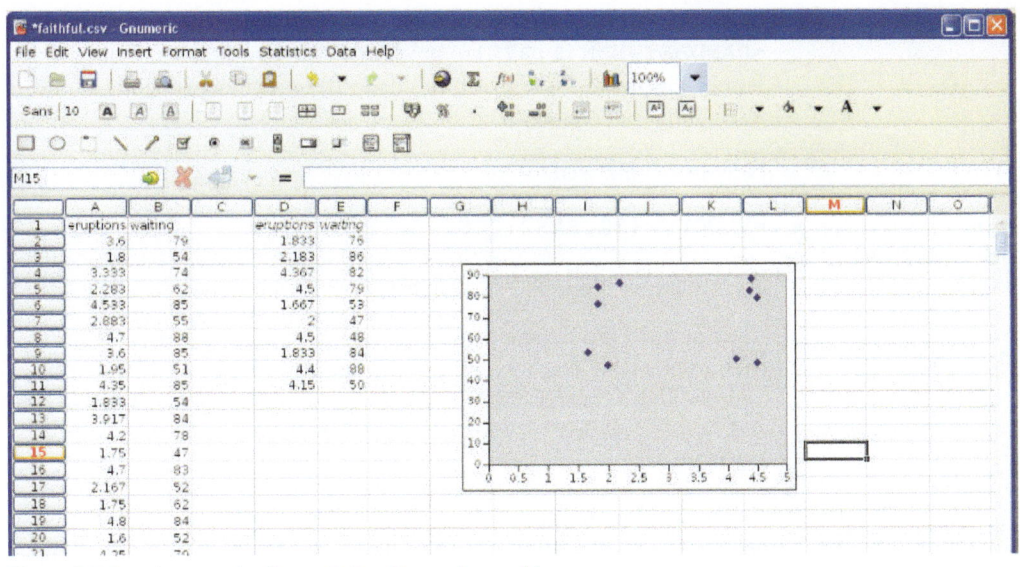

Figure 8.7 Inserting a graph of a sample in a Gnumeric spreadsheet

simple to demonstrate how a small sample can easily result in contradictory interpretations if the fledgling statistician is unwary.

8.5 Doing the same thing with Excel

Taking a random sample from a bivariate data set is straightforward in Excel, using the Data Analysis Tool, and it follows mostly the same steps as the Gnumeric walkthrough above. The problem for our purposes is that, unlike Gnumeric, Excel returns values as output not formulae, so we have to create the formulae for ourselves.

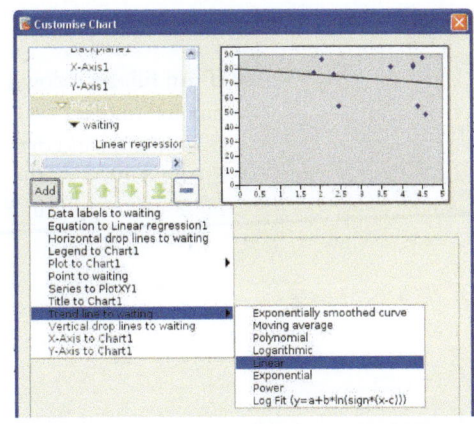

Figure 8.8 The graph settings window in Gnumeric

Put your data into columns B and C of an Excel sheet, then put in column A an index from 1 to n so that each data pair has an integer associated with it.

In our example using the Old Faithful data, the top few rows should look like Figure 8.9.

In order to take a sample of size 10 we need to generate 10 random numbers between 1 and 272 (the size of the data set) and then use the **VLOOKUP** command to look up in the first three columns the pair that is indexed by each of the random numbers.

	A	B	C	D
1		eruptions	waiting	
2	1	3.6	79	
3	2	1.8	54	
4	3	3.333	74	
5	4	2.283	62	
6	5	4.533	85	
7	6	2.883	55	
8	7	4.7	88	
9	8	3.6	85	

Figure 8.9 Arranging bivariate data in an Excel spreadsheet

101

In column E, from cell 2 down to 11 write the formula
`=RANDBETWEEN(1,272)` to produce 10 random numbers from
1 to 272 inclusive. Note that this is done 10 times independently, so this
is sampling *with* replacement.

Now, in cell H2 write

`=VLOOKUP(E2,$A$2:$C$273,2)`

We will go through this line in detail.

The VLOOKUP command looks up the value in cell E2 in the leftmost
column of a table that we define as the range from A2 to C273, so it
finds the index that has been randomly generated in cell E2. It then looks
at the corresponding value in column 2 ('eruptions') and writes it into
cell H2.

To get the corresponding value of 'waiting' into cell I2 you need the very
similar command `=VLOOKUP(E2,$A$2:$C$273,3)`. This looks up
the same row in the table, but returns the value in column 3 instead of
column 2.

If you fill these down from cells H2 and I2 as far as H11 and I11 (by
highlighting H2 and I2, then grabbing the box in the bottom right corner
of the frame and dragging down), you should end up with a sheet like
Figure 8.10, with the original data in columns B and C, the 10 random
numbers in column E, and the sampled pairs in columns H and I.

	A	B	C	D	E	F	G	H	I
1		eruptions	waiting					eruptions	waiting
2	1	3.6	79		124			1.967	56
3	2	1.8	54		61			2.233	59
4	3	3.333	74		167			2.367	63
5	4	2.283	62		254			4.5	73
6	5	4.533	85		207			4.367	77
7	6	2.883	55		160			3.967	89
8	7	4.7	88		87			3.95	76
9	8	3.6	85		46			3.317	83
10	9	1.95	51		260			4.283	79
11	10	4.35	85		200			4.667	78
12	11	1.833	54						
13	12	3.917	84						
14	13	4.2	78						
15	14	1.75	47						

Figure 8.10 Taking a random sample of size 10 in an Excel spreadsheet

Now it is plain sailing. To create a scatter graph, highlight columns H and
I, then click 'Insert' and choose 'Scatter' from the 'Charts' box. You can
then refresh the sample by pressing F9 and watch the sampling variation
happen live. If you want to add a trend line, simply right-click on a data
point and choose 'Add Trend line...'. This is a bit of a pain to set up in

Excel, but once you have got your head around the **VLOOKUP** command it is not too bad. What you don't have is the ability to very easily change the size of the sample, the number of samples and so forth. If you are happy with writing macros then this is perfectly doable, but you could just download Gnumeric instead …

8.6 Summary

In this chapter we have shown how one particular activity (repeated sampling) can be implemented using technology as a tool to offload the computation required to produce multiple samples and graphical representations. This allows students to see and experience variability in a way which is simply not possible with pen-and-paper methods. We hope that this might be the starting point for hesitant readers to explore the possibilities of using technology in their own teaching.

This chapter has given a step-by-step route to creating one particular activity. In the following chapters we will take a look back at the main themes of the book with suggested activities for the classroom, some using a little technology, others just manipulatives and pen-and-paper. The purpose of these activities is to provide a starting point for the introduction of a more holistic approach to the delivery of statistics, focusing on the experiences that students should have in order to develop statistical literacy.

PART 2 Activities

Introduction to the activities

The chapters in Part 1 covered the key points to consider when developing statistics lessons and activities. Statistical literacy cannot be taught in a single lesson and should be treated as a developmental process. Students learn new techniques by using appropriate data, always remembering that data are numbers in context, not simply lists of numbers. They then require multiple opportunities to select and use these new skills alongside the other tools in their statistical toolkit.

Activities should aim to create opportunities for students to focus on one or more of the following elements of statistical literacy while learning to apply the standard techniques and skills required by the curriculum they are following.

- Using the complete statistical cycle
- Describing data in context
- Inventing and refining graphical representations
- Carrying out exploratory data analysis
- Identifying sources of variation
- Searching data for signals in the form of averages
- Considering data as an aggregate and as individual points
- Summarising distributions
- Making and justifying inferences
- Simulating and modelling data

Each activity in Part 2 is designed to reflect the content of a chapter from Part 1, and it is recommended that the related chapter is read before trying the activities with a class in order to get the most out of them. These activities are designed as exemplars and can be adapted as necessary to suit the needs of specific students. Having read the associated chapter, teachers should be well placed to adapt and enhance the activities to meet their specific requirements.

The set of activities does not constitute complete curriculum coverage, but the ideas can be adapted to cover many different statistics techniques. Often students will benefit from returning to similar activities over an extended period by keeping a similar structure but using a different data set to prompt different questions, representations and conclusions.

PART 2 Activities

The world is full of data and most students love to share their opinions. Building lessons and activities by finding data based on the issues that are important to your classes can be incredibly motivating for both students and teachers and may help students engage with complex statistical ideas beyond what would normally be expected of them. The key message is: use data to help students engage in constructive argument and, above all, have fun!

Chapter 9

Activity for Chapter 2: The statistical cycle

Finding fault

Planning for the classroom

This activity is suitable for all levels, with more advanced students able to analyse the data using more sophisticated skills such as standard deviation to explore beyond the initial problem. The main activity can be completed using only simple averages.

In the classroom teachers need to balance the requirement of students to practise specific technical skills with the need to experience rich forms of inquiry that address the complete statistical cycle. In this activity, students are provided with opportunities to repeatedly calculate the mean within a framework that allows them to use these calculations to meaningfully explore problems and argue in support of their conclusions.

Equipment and resources

Before the lesson each group (pairs or threes suggested) will need:

- access to the data set *Quality control data*
- the *Finding fault* worksheet.

Suggested approach

The goal of this task is for students to identify how the data could be used to pinpoint the time at which the machine began to break down and hence identify the fault that happened shortly before this time and recommend a solution.

The approach to this task can be modified depending on the prior experience of students in working with the complete statistical cycle and calculation of the mean. For less experienced students, teachers could model the initial parts of the cycle, providing additional guidance and direction towards calculating the mean value of each small sample.

Initially students should be given time to consider the problem in context, the following prompt questions may help.

> **Discussion questions**
> - Why can't the maintenance team check all the possible problems?
> - How might the quality control data be useful?
> - What effect might you see in the data as the machine begins to go wrong?
> - What could you do to the data to make it easier to see any effects?
> - Are there any calculations you could do to help spot patterns in the data?
> - Are there any graphs you could draw to help spot patterns in the data?

The purpose of this task is to provide an opportunity for students to practise calculating the mean with a purpose; however, due to the large number of calculations, a spreadsheet may be an appropriate tool.

Students could be encouraged to produce spreadsheet formulae for the calculations, rather than using inbuilt average functions, and to record them along with example calculations as part of their working. Alternatively, the calculations could be shared amongst the students, with the teacher collating the answers for the whole class.

Once students have a complete set of mean values, they should be encouraged to graph the data and construct their argument for which is the correct repair for the team to perform.

An alternative approach, using technology, would be a simple piece of exploratory data analysis using Excel's conditional formatting. Simply highlight all of the data in the sheet and apply a standard colour scale to the data. It is immediately obvious by sight the point at which the weights begin to vary significantly, but you might like to encourage students to confirm their conclusion using summary values or other graphical representations.

An important part of the statistical process is to reflect on the conclusions drawn, as well as the methods employed. Students should be encouraged to share their graphs and discuss the decisions they made. Below are some questions to help guide this discussion.

> **Discussion questions**
> - How confident are you that you have suggested the correct course of action to fix the machine?
> - Which graphical representation shows the data most clearly?
> - Why did you decide on the scale you have chosen?
> - Could your graph be improved to show the data more clearly?
> - Could you have approached this problem using different techniques?

Additional discussion points

To enrich the context further, teachers may wish to encourage exploration and discussion of the issues involved in the context, such as:

- whether supermarkets should sell vegetables and fruit in standardised sizes and shapes
- the supply chain of food from field to plate
- the advantages and disadvantages of 'just in time' stock control.

Finding fault worksheet

A potato sorting machine has broken down!

Supermarkets like their produce to be a consistent size and shape, so the potato sorting machine selects all the potatoes within a certain weight range to be sent to the supermarkets to be sold in bags; any potatoes outside of this range are sold to be used as ingredients in ready meals instead.

At 13:45 yesterday the potato sorter shut down unexpectedly.

The machinery is large and incredibly complicated, so before it can be repaired the maintenance team need to work out what caused the problem and identify which part to fix.

The factory manager has produced a list of unusual things that happened over the previous 48 hours that she thinks may have caused the fault in the machinery.

Time	Event	Solution
Day 1 06:00	During a storm, lightning struck the factory building.	Check and replace the fuses in the main control room.
Day 1 11:00	Susan from accounts opened an email that caused the security software to create an alert although no virus was initially found.	Run a full virus scan of the system to check for malicious software.
Day 1 15:15	Some new parts were fitted to the machine by Sandra, the engineer.	Reinstall the older parts and check the new ones for faults.
Day 1 21:50	The main computer briefly logged a sensor reporting a high temperature in part of the machine, but it quickly returned to normal.	Check the sensor is functioning properly and replace if not.
Day 2 03:20	The burglar alarm went off although no intruder was found by Jon, the security guard.	Check through the CCTV images for an intruder in case someone damaged the machine deliberately.
Day 2 07:20	A storm caused a power cut locally and the factory backup generators were providing electricity to the factory up until the production line broke down.	Service the generators and reset the machine.
Day 2 13:00	Malcolm, the cleaner, knocked over a bucket causing water to leak into part of the machine.	Replace the parts that got wet.

Each solution will take several hours to resolve and the maintenance team can only address one thing at a time. While the machine is not running, the entire supply chain is disrupted. Lorries of produce keep arriving with more potatoes for sorting, and orders must be fulfilled in order for the supermarkets to keep their shelves stocked. It is essential to identify the cause of the problem before deciding what to do.

Every hour, a sample of between 8 and 12 sorted potatoes are automatically weighed for quality control, and the manager thinks that this data might be useful in spotting what has gone wrong. This data has been provided to the maintenance team.

Can you work out how to fix the machine?

Quality control data

Sample potatoes (mass in grams)

		Potato 1	Potato 2	Potato 3	Potato 4	Potato 5	Potato 6	Potato 7	Potato 8	Potato 9	Potato 10	Potato 11	Potato 12
Day 1	01:00	123.4	123.7	122.2	123.6	123.8	123.8	122.5	123.5	124.2	123.1		
	02:00	123.4	122.7	123.6	123.0	123.5	121.9	122.1	122.8	123.3	123.7		
	03:00	123.3	124.1	123.3	122.5	122.5	123.6	123.0	122.9	122.8	123.0		
	04:00	123.0	124.0	123.2	123.0	123.6	122.3	123.2	122.4	123.1	123.4		
	05:00	123.8	123.1	122.9	123.2	122.8	122.6	123.4	123.6	123.2	123.9		
	06:00	123.1	122.6	123.0	122.9	122.9	121.8	123.3	123.1	123.7	123.2		
	07:00	123.9	123.6	122.4	123.6	122.4	122.7	123.4	123.2	124.0	123.4		
	08:00	123.4	123.3	122.5	123.1	122.3	122.4	122.5	123.3	123.5	123.1		
	09:00	123.4	123.1	122.7	122.5	121.8	123.4	123.7	123.5	123.7			
	10:00	123.0	122.7	123.4	123.7	123.6	122.3	122.3	122.6	123.0	123.5	124.0	
	11:00	123.2	123.0	124.0	122.7	123.2	123.8	123.4	123.7				
	12:00	123.1	122.6	123.8	123.2	123.0	122.2	122.7	123.1				
	13:00	123.7	122.9	122.8	123.2	122.9	123.0	123.9	123.0	122.4	123.1	124.1	
	14:00	122.7	123.2	122.2	124.0	123.8	122.5	122.6	123.1	122.9	123.8	122.8	123.4
	15:00	123.1	123.5	123.6	122.4	124.0	123.0	122.6	123.2	122.8			
	16:00	123.6	123.8	123.8	123.8	125.0	124.4	123.0	124.3	124.2	123.5		
	17:00	124.3	124.3	123.8	123.8	124.1	123.4	123.0	124.0	123.9			
	18:00	124.6	123.8	123.8	123.5	123.7	124.0	124.4	124.2	124.4	124.2	123.4	
	19:00	124.4	125.3	123.4	123.7	123.0	123.8	123.2	124.1	124.0	123.8		
	20:00	123.1	123.8	124.2	123.0	124.1	123.9	123.4	123.8	123.8	123.5		
	21:00	123.5	123.8	124.9	124.5	124.4	123.1	124.3	124.2	122.6			
	22:00	124.7	123.9	123.4	124.3	123.5	123.9	124.5	123.1	123.7	124.2	124.1	
	23:00	123.4	124.5	124.2	124.4	123.7	124.4	124.4	123.5	124.3	124.4		

PART 2 Activities

111

Sample potatoes (mass in grams)

		Potato 1	Potato 2	Potato 3	Potato 4	Potato 5	Potato 6	Potato 7	Potato 8	Potato 9	Potato 10	Potato 11	Potato 12
Day 2	00:00	123.3	123.8	124.8	123.6	124.1	123.6	123.5	124.0	122.6	123.3	124.4	
	01:00	124.2	124.5	124.7	124.2	123.1	122.9	124.7	124.2	124.5	124.0		
	02:00	124.9	124.1	124.0	123.7	124.2	124.3	124.2	124.4	123.8	124.0	123.1	
	03:00	124.1	124.1	124.0	124.2	124.6	123.0	124.1	123.9				
	04:00	125.2	124.3	124.5	123.9	123.4	123.9	123.7	123.8	124.5			
	05:00	123.1	124.3	123.6	123.9	123.5	123.7	124.0	124.0	124.0	123.9	124.2	123.8
	06:00	124.7	124.0	123.6	124.3	124.4	123.4	123.1	123.3	124.6	124.3	124.6	124.1
	07:00	124.1	124.1	124.4	123.6	124.7	124.5	123.7	123.2	123.9	123.2		
	08:00	124.6	124.0	123.0	123.7	124.4	123.9	123.7	123.2	124.0	123.8		
	09:00	126.3	118.7	119.4	120.9	123.9	121.0	121.3	122.9	121.9			
	10:00	120.5	123.3	121.4	121.0	120.3	126.2	124.4	124.2				
	11:00	121.8	124.9	122.9	122.9	124.0	124.3	122.2	124.2	120.4	126.2		
	12:00	122.1	123.9	123.9	120.6	123.8	124.8	126.4	124.3				
	13:00	122.8	128.2	121.7	125.4	122.4	118.9	123.5	120.1	122.7	124.0		

(continued)

The number of data points in each sample ranges between eight and ten; however, the first few are always ten for ease of calculation.

Chapter 10

Activity for Chapter 3: Exploratory data analysis

Using Gapminder

Planning for the classroom

This activity is suitable for all levels. It helps students become familiar with reading data from scatter graphs and seeks to avoid the technical difficulties of working with exploratory tools. It allows students to practise drawing informal inference from dynamic data visualisations and to experience the potential depth and breadth of exploration using *Trendalyzer*. The activity could be run as a whole-class activity, but students will only gain the direct experience of following their own pathway if it is undertaken in small groups.

This activity could form one lesson, or could easily be developed into a small project.

Equipment and resources

Each group (pairs or threes suggested) will need:

- access to the Trendalyzer tool (www.gapminder.org/tools/)
- the *Using Gapminder* worksheet.

Suggested approach

This activity is in part an introduction to using the Trendalyzer tool and the particular style of visualisation, but it is mostly about introducing students to using visualisation techniques to identify patterns of interest from large sets of data. It provides a good context for asking questions, creating and discussing hypotheses, and thinking carefully about what other data one might need to support the hypotheses. There is plenty of interesting behaviour to look for and attempt to explain, possibly by doing a little historical research. If your setting allows for some cross-curricular delivery, investigating Gapminder data in collaboration with, for example, geography or history teachers would be very constructive.

The activity is described below as if carried out in class, but much of it, particularly the latter parts of exploratory data analysis, would work very well as a flipped resource or small project, if your students have access to the internet while outside the classroom.

Set-up

For Activity 1, you should set it up in advance so that students do not see the use of bubbles in the first instance. To decide which data to show on the axis, click on 'Options' and select 'X AND Y'. Set up the x-axis as 'Urban population (% of total)' and the y-axis as 'Life expectancy (years)'; also choose the time to be 1960 by adjusting the slider below the graph. Go back into 'Options', then 'SIZE', and move the slider down so that all the bubbles appear as small coloured dots of about the same size.

This creates a static scatter diagram; the other two dimensions will come later, but it is helpful to keep them out of the way for the initial questions about overall trends and predictions. Note that the intention is not to trick your students by withholding information but to give them an opportunity to test and hone their intuition, and to begin their analysis of the data on two variables before throwing in too much detail. If you like, you can play more with this initial set-up, giving certain countries more emphasis (e.g. by continent, your own country).

Activity 1 – getting to know the context and looking at general trends

Give the following prompts to the whole class, and allow students some time to think and discuss in small groups; but keep in mind that the main point of the activity is familiarisation with the context and the measures involved.

- What is the overall trend? Do you think that the scatter graph shows an association between the variables or not?
- Note that the time is set at 1960. What would you predict is going to happen over time?
- Where do you think the countries with large populations are?

Note that the final question is tricky; the highest urban populations are in tiny countries like Singapore, and students may find themselves with little intuition about countries in 1960.

Before starting the animation, change 'Size' to show the total populations. How does the resulting graph match students' intuition?

Activity 2 – changes over time and forming hypotheses

This activity is essentially one of exploratory data analysis. Students will use the data visualisation to investigate and explore the data, form hypotheses and test them against historical fact, or simply use features of the data as a motivation for further research. If you have the equipment for students to do this activity in small groups, that is far better since they can then pause or rerun the animation and will be able to pick out the interesting countries for themselves. Give each group of students the *Using Gapminder* worksheet. This shows them how to set up Gapminder and provides prompts to get them going. Leave enough time at the end for students to present a few ideas to the whole class.

Although you can run this as a whole-class activity using the prompts below, students will miss out on the direct experience of working with and exploring the data. It is important to allow students to lead the exploration as much as possible when working as a class, allowing them to formulate their own questions.

Start the animation. Simply watch it through once. Then ask your students to watch it again and note down any general trends or movements, keeping an eye out for strange behaviour. Now watch it a third time and ask students to shout out, or write down, when they spot something strange to investigate further.

Discussion questions
- Which countries have you picked out as doing something interesting?
- Can you give any explanation for this behaviour? You may not know anything about the given country, but you might be able to come up with a possible cause.
- Use the internet and a search engine to investigate the country around the year in which you have found the interesting behaviour. Do you have an explanation now?

NB You may find the trace function useful, particularly if picking out a country at the front of the class, or comparing two countries.

Interesting countries to watch for and some possible follow-up

- Add the trace to Cambodia or Rwanda (or Kuwait or Ethopia) to see a sudden dramatic drop in life expectancy but an opposite effect on urban population. Does a drop/rise in urban proportion *necessarily* mean a drop/rise in urban population?
- Haiti and Indonesia suffered large population losses due to natural disasters in this period. Do these show up in the data?

Can you think of a measure to replace one of the axes that might show a bigger effect? This is a great opportunity to talk about subpopulations. Child mortality and median age are good choices of variables to show robustness.

- Watch the Channel Islands for a country with a general decrease in urban population over the time period. It is much harder to find a direct cause in this case, maybe it is the small islands? Try looking at Jamaica, Maldives and similar islands.

- Observe Botswana for a general increase in urbanisation but sudden drop in life expectancy due to HIV prevalence. Try changing 'Size' of the data points to show HIV percentage to see this effect.

- Japan is an example of a country with a fairly steady growth until the end where there is a period of rapid urbanisation.

- Follow the United Kingdom to see an example of very little change in urbanisation. Why might this be the case? When might this have happened in the past?

Some other indicators to try in place of urban population share on the horizontal axis (but keeping life expectancy on the vertical axis) might be:

- Military expenditure: what happened in Kuwait in 1991?

- Internet users: Morocco and Iraq are interesting – why are there lots of step changes horizontally, but steady growth vertically?

- Child mortality rate: shows negative association with life expectancy and explains the difference seen above between the Haiti earthquake in 2010 and the tsunami in Indonesia in 2004 (i.e. in Haiti there was a much greater effect on child mortality, which brings the life expectancy down in a way that the more general loss of life did not do in Indonesia). Can you spot Live Aid?

Using Gapminder worksheet

In this activity you are going to be using the Trendalyzer software to explore data.

1. Open a browser window and go to www.gapminder.org, then click on 'Gapminder tools' in the top menu.

2. To decide which data to show on the axes, click on 'Options' and select 'X AND Y', as ringed in Figure 10.1.

    Set up the *x*-axis as 'Urban population share' by clicking on the drop-down menu and selecting 'Population', then 'Urbanization', then 'Urban population (% of total)'.

    Leave the *y*-axis as 'Life expectancy (years)', which it should be set on by default.

    Before you start, make sure that you understand what these measures mean. Can you work out why the data points are bubbles of different sizes and colours?

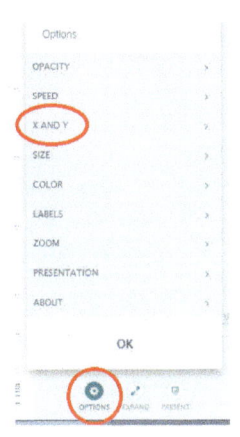

Figure 10.1
The Trendalyzer Options menu

3. Either drag the time back to 1960 using the slider at the bottom of the screen, or simply press the play button in the bottom left to start the animation.

4. Watch the animation several times and note down any general trends or movements that you notice. Keep an eye out for strange behaviour.

    You can pause and restart the animation at any point; just drag the slider at the bottom to move to a particular year.

    - Record the names of any countries that you picked out as doing something interesting.
      (You can find the name of each country, as well as the values of the two measures plotted, by hovering the mouse over the data point.)
    - Are there countries that seem to behave in similar ways?
    - Can you give any explanation for the behaviour you observed? You may not know anything about the given country, but you might be able to come up with a possible cause.

    Using the internet and a search engine, investigate one country around the year in which you have found the interesting behaviour. Do you have an explanation for the behaviour now? If you want to watch a single country progress through time, you might find the

'Trace' function useful. Click in the box next to country's name on the right-hand side and as the animation plays the software will join up the dots for that country.

5   Write down your conclusions, listing any general trends and patterns spotted, along with any behaviour you have noted which is different from the norm and any explanations you think you may have found.

6   You can continue your investigation using any of the measures in the data set by clicking on 'Options', 'X AND Y' and choosing different variables to plot along each axis. Try just changing one thing at a time and note down any interesting trends or strange behaviour.

Chapter 11

Activities for Chapter 4: Simulation

Activity 1: Completing collections

Planning for the classroom

This activity is accessible for all levels. It is designed to help students understand the power of simulation to deal with an analytically complex situation and also to provide the opportunity to hone their own intuition and understanding of statistics without getting bogged down in calculation. Older students are more likely to have the IT skills needed to simulate the scenario with technology, but we will consider only the physical simulation here.

Equipment and resources

Each student (or group of two or three students) will need:

- six different-coloured multilink cubes or similar objects
- a bag to put them in.

Suggested approach

A nice example of a distribution which is slightly counter-intuitive, but which students will know, is that arising from trying to collect a set of toys which you find distributed at random in cereal packets, chocolate eggs, fast food meals and so on. For lower secondary students, this distribution is not approachable theoretically, so how could we approach it with a simulation?

One way to simulate it physically is with a randomisation device which can be tailored to change probabilities as toys are 'found', for example a bag containing beads or multilink cubes.

Start by introducing the context and the question 'How many packs am I going to need to open in order to get a full set?' Note that the question allows for the possibility that the answer is a given number of packets; leave the idea that it might take a short or long time to come out during discussion. Once it has, it is worth also asking 'What are the smallest and largest numbers of packs that I might need to buy?' Again, the question leaves the potentially infinite nature of the problem to come out in discussion.

Once armed with their intuition, preferably with predictions written down for later, students can attempt the physical simulation using the following algorithm, keeping a tally of the number of draws required. You should discuss with them how this models the situation. They don't need to understand the probabilities to understand the situation since the model and the situation we are modelling are very close. What modelling assumptions have been made? Will these affect the validity of our outcome?

> STEP 1 Start with six different-coloured cubes in the bag.
> STEP 2 Draw out one cube. Write down its colour, add a tally mark for it and replace it in the bag.
> STEP 3 Draw out one cube. If it is the same colour as one on your list, add a tally mark, replace it in the bag and repeat STEP 3; otherwise go to STEP 4.
> STEP 4 Write down the colour of the cube just drawn on your list and add a tally mark. If your list has six colours then STOP; otherwise replace the cube in the bag and return to STEP 3.

At the end students should have written down the number of draws it took to collect all six colours. Assuming that this takes a similar amount of time for each student or group of students, take a quick poll, perhaps collating results at the front of the class, and ask students whether or not their result matched their prediction.

> **Discussion questions**
> - Does your result match your prediction?
> - If not, is your prediction wrong or is your data an outlier?
> - How can you tell?

A conclusion cannot be drawn from a single data point, but the class will have generated a decent amount of data between them, particularly if you have time to repeat the activity multiple times.

The expected value is 14.7 (why?), so you might find some students taking a while to get all six colours and they might need stopping if they get really unlucky!

There are various simulations of this type of scenario available on the internet, for example the one at https://mste.illinois.edu/reese/cereal/

cereal.php, which has some nice animal toys to look for. If you want to generate data more quickly, then this would be an easy-to-use option.

To answer the question about how much you have to spend on cereal to get all the toys, students will need to make some assumptions, the most obvious being the cost of a single box of cereal. It may be useful, if you have enough classroom time available, to encourage an extended discussion of this assumption as the precise brand of cereal is not given.

Discussion questions
- Should you pick a specific cereal and use that price or should you look at a range of cereals and choose a representative price?
- Is it reasonable to estimate a price without looking at individual cereal costs?
- Does the choice of store affect the cost?

Other assumptions that may be worth discussing relate to the nature of production methods and how accurate the simulation model is.

Discussion questions
- Is it reasonable to assume that every toy is equally likely to be found in a random selection of packs and are the probabilities independent?
- Does batch production have an impact? For example, is it likely that all the boxes of that type of cereal in a supermarket on a particular day contain the same toy as they all came from the same batch in the factory?

How far you want to go with these discussions is dependent on the engagement of students in the task and the time you have available.

'How much do the toys cost?' worksheet

Every box of a particular breakfast cereal contains a random free toy, with six different toys in total to collect.

Question: How much would a family have to spend on cereal to get all the toys?

You are going to work this out by running a simulation to collect data. You will need:

- six objects to represent the toys, such as different-coloured multilink cubes
- a bag or pot to draw them from.

Simulation method:

Use this algorithm, keeping a tally of the number of each colour drawn.

> STEP 1 Start with six different-coloured cubes in the bag.
> STEP 2 Draw out one cube. Write down its colour, add a tally mark for it and replace it in the bag.
> STEP 3 Draw out one cube. If it is the same colour as one on your list, add a tally mark, replace it in the bag and repeat STEP 3; otherwise go to STEP 4.
> STEP 4 Write down the colour of the cube just drawn on your list and add a tally mark. If your list has six colours then STOP; otherwise replace the cube in the bag and return to STEP 3.

The total of your tallies represents the total number of cereal boxes opened.

Part 1

Based on the results of your simulation:

1. Using your data, how much do you think it would cost to get all six toys?

2. How certain are you that your data is reliable?

3. How could you improve the reliability?

Part 2

Share your total number of boxes opened with the rest of the class and create a single graph/chart showing this data.

1. Do you still agree with your answers from Part 1?

2. What does the new graph tell you about the cost of getting all six toys?

Extension

The cereal company decides to make one of the toys half as likely to be found as the other five.

1. How could you adjust your simulation to take this into account?

2. What effect does this decision have on the cost of getting all six toys?

Activity 2: Long-term stability vs short-term variation (Law of large numbers)

Planning for the classroom

This activity is suitable for all levels, though more advanced students will need less help to create the spreadsheet for themselves. It would be best if students work in small groups, to allow discussion of what they are seeing at a detailed level. Care is needed to ensure that students do not simply generate a very large number of graphs and pick out the weird-looking ones.

This spreadsheet-based simulation activity is intended to allow students to experience the law of large numbers in a relatively simple context, that of repeatedly flipping a fair coin. This will include thinking about the conflict between short-term variability and long-term stability, as well as the damping effect of recording cumulative data. The extension task, with hidden probability of heads, is a rich source for informal discussions about prediction and confidence intervals.

Equipment and resources

You can give the spreadsheet to students with formulae in place, or they can create it for themselves from scratch. The latter task is not demanding, particularly with step-by-step guidance (provided at the end of the activity), and it has the far better pedagogical result that students understand the simulation they are experiencing.

Simply giving students the version with 1500 coin flips does create a barrier to their understanding what the graph is showing them, so it is more sensible to take a growing samples approach, starting with a small sample and building up.

Suggested approach

This section assumes that you are starting with a pre-constructed spreadsheet with separate tabs for $n = 15$, 150 and 1500 as described below.

Understanding the construct

Start with the tab labelled $n = 15$. Students will see a short list of numbers and a line graph.

The first step is to help students get their heads around what is being displayed. The spreadsheet generates either 1 or 0 to represent heads or tails, and then calculates the proportion of flips which have come up heads. This simulates a person flipping a coin, keeping track of the results and a running proportion. If you prefer, you can get students to carry out this first stage by hand.

The beauty of the spreadsheet is that you can very quickly resample by pressing F9 (assuming you are using Excel). As your students press F9 repeatedly, they will get new samples and be able to see very quickly on the line graph what is going on. They can save interesting-looking graphs for later by copying the box to the clipboard and pasting it as an image into a document.

> **Discussion questions**
> - What counts as 'interesting'?
> - What does the variability tell us about the coin?
> - What does the variability tell us about the probability of a head?
> - What do you think will happen as we increase the value of n?

There is a good chance that any graphs selected as interesting enough to keep are outliers. This may be a good opportunity to explore the idea of selection bias and/or confirmation bias.

Increasing n

The next step is to increase n. The spreadsheet has a tab for each of $n = 150$ and $n = 1500$. Make sure that students have predicted what they think might happen. It is tempting to say the graph will smooth out, which it does. It is also tempting to say that there will be less variation, and/or that the long-term value will be closer to 0.5 than in the $n = 15$ case. It's easy to show that is not true.

> **Discussion questions**
> - Does the proportion of heads either decrease or increase all the time?
> - Once it has got close to 0.5 does it stay there?
> - Can you guarantee that the proportion will be within a given range at $n = 150$?

The worksheet is designed to support students' initial engagement with the simulation activity. The questions, particularly in Part 2, are designed to be prompts for further discussion, and it is up to the teacher whether to insist on students keeping a written record. A possible approach is to ask students to write a report at the end of the simulation activity, considering the prompts, in order to encourage a more reflective narrative rather than single-sentence answers.

Law of large numbers worksheet

Part 1

You will need a coin. Each time you flip the coin, score 1 for a head or score 0 for a tail.

1. Flip the coin 5 times, write down the total score.
2. If you got 10 heads, what would the score be?
3. If you got 10 tails, what would the score be?
4. After 10 coin flips, what would the mean score per flip be if you got
   a) 10 heads   b) 10 tails   c) 5 heads   d) 3 tails?
5. Regardless of the number of coin flips, what does the mean score per flip represent?

Part 2

Using the computer simulation, investigate the mean score of a series of 15 coin flips. Press F9 to simulate 15 coin flips and analyse the different graphs produced. Save any interesting graphs into a document by copying and pasting each graph as a picture.

1. What do the values on the x-axis of the graph represent?
2. What do the values on the y-axis of the graph represent?
3. What made you decide that your chosen graphs were 'interesting'?
4. What does the variability tell you about the coin?
5. What does the variability tell you about the probability of a head?
6. What do you think will happen if you increase the number of coin flips beyond 15?

Now repeat the simulation with 150 coin flips and 1500 coin flips.

1. Does the proportion of heads either decrease or increase all the time?
2. Once the proportion has got close to 0.5 does it stay there?
3. Can you guarantee that the proportion will be within a given range at $n = 150$?

Step-by-step instructions for creating the spreadsheet

Setting up the *n* = 15 tab

- In cell A1 write `=RANDBETWEEN(0,1)`, which generates a random number, either 0 or 1, with equal probability.

- Highlight the cell, then use the 'fill handle' in the bottom right of the cell to drag the formula down to row 15.

- Now we need to set up the sample averages. In cell B1 write `=AVERAGE($A$1:A1)`, then fill down to cell B15 as before. You should find that the $A$1 bit stays the same down the column (this is what the $ sign does), but the A1 changes with each cell. In cell B15 it should read `=AVERAGE($A$1:A15)`. For each cell in the B column you should now see the average of the A column values from that cell upwards.

- As you did all of this, you will have noticed the spreadsheet updating itself all the time. Every time you make a change, it will resample the values in column A. You can also make this happen by pressing F9.

- Now insert a line graph. This may, or may not, create the graph you are after automatically. You should right-click on the graph, choose 'Select Data' and make sure that it is displaying only column B under 'Legend Entries (Series)'. Then click 'OK', to leave.

- As you press F9 the spreadsheet will generate a new sample of 15 coin flips and draw the graph of the sample averages.

Setting up the *n* = 150 and *n* = 1500 tabs

If you right-click on the tab for the sheet you are working in, choose 'Move or Copy…', tick the 'Create a copy' box and click 'OK', you will get a copy of the sheet (unsurprisingly!).

Use the 'fill handle' again to drag the formulae in column A and column B down to cell 150 (or 1500). All you need to do then is right-click on the graph, choose 'Select Data' and make sure that it is displaying the range from B1 to B150 (or B1500) for the data.

Chapter 12

Activities for Chapter 5: Sampling and variation

Activity 1: Growing samples

Planning for the classroom

This activity is suitable for all levels. Its aim is to help students get a feel for how the size of a sample affects the variability of data, get used to drawing meta-representations (quick sketches to represent the general shape) of data, and develop a descriptive language around the aggregate properties of data.

Equipment and resources

You will need access to a data set, or you could conduct a census of students in the school in order to collect data which can be explored over the course of the year. A good data set for dotplots will be fairly small and discrete and in a familiar context. The internet is replete with interesting data sets that can be downloaded for use with students; for example, www.mei.org.uk/data-sets hosts a number of their own along with links to other sites. The 'Wingspan' field in the 'Blackbirds' data set is a good starting point. This is a discussion-based activity led by the teacher where students draw graphs of increasing samples before discussing their results.

Growing samples worksheet

While you could produce printed prompt questions with spaces for answers, because this is a teacher-led activity it is not necessary and the focus of the lesson should be on extensive student discussion rather than filling in a worksheet.

Suggested approach

First have each student randomly sample a single piece of data; this could be done using a spreadsheet or R if you have the technology available, or by pen and paper methods.

Ask students what they can tell from the single data point. This is likely to be descriptive ('Mine is 152'), and where students have some idea of the context they may give a value judgement as to whether this is an underestimate or overestimate.

Ask students to collect four more pieces of data to bring their total to five and draw a dotplot to represent the data (see Figure 12.1).

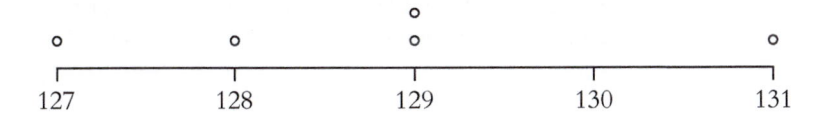

Figure 12.1 Dotplot showing the wingspan of five blackbirds

Ask students to write a description of their graphs before sharing their dotplots with the rest of the class, either by getting some students to draw their graphs on the board or by using a web cam if available. (Free software can be downloaded that allows you to use a smartphone as a web cam but will need installing on both PC and phone in advance.) The following questions can then be discussed.

> **Discussion questions**
> - How do the graphs differ?
> - Are there any similarities?
> - What does your graph tell you> Can you describe it?
> - Are any of the graphs identical?

Now ask students to pool their data with a partner, and draw a representation of the 10 data points.

Ask students to write a description of the new graphs and again share them with the rest of the class.

> **Discussion questions**
> - How have the graphs changed?
> - Has any information been lost?
> - Do any new features stand out?
> - How does each graph compare to other pairs' graphs?

Repeat the process again, with four students pooling their data, then eight. At this point students will have 40 data points to plot and a stable picture should be emerging with consensus between groups.

Show the students a graph of the entire population and discuss how similar it is to their representations.

Discussion questions
- What similarities do the sample graphs have to the population?
- What differences are there?
- How big was the sample before it began to look like the population graph?
- Do any of the larger samples not look like the population graph?
- What have you learnt about sampling?

Other avenues to explore

For a group of students familiar with averages and range, they could be asked to calculate these statistics as each sample grows and include reference to these values in their discussion. Be careful not to allow students to focus on the measures of central tendency at the expense of the variability.

This activity could be repeated with different data sets, including continuous data, with the dotplot drawn on a continuous scale.

If students have access to a computer with the data set, they could do this activity independently with a worksheet, guided by prompt questions to answer at each stage. It should be noted that the discussion of other students' graphs with the same-sized sample from the same population is an important aspect of this activity.

PART 2 Activities

Activity 2: Distribution

Planning for the classroom

The aim of this activity, which is suitable for all levels, is to help students explore the relationship between the five-number summary, boxplots, the location of data and the shape of distributions. Before beginning the activity students should be introduced to the median as the middle value. They do not need to have worked with boxplots before, but they will require more support if this is the situation. The activity can be undertaken individually or in pairs or small groups.

Equipment and resources

This activity can be approached in a variety of ways, using ICT, pencil and paper, or multilink cubes or similar. Details of each approach are given in the Suggested approaches section below. Students will need a copy of the *Distribution* worksheet, which asks them to create graphical representations of distributions from the boxplots.

Suggested approaches

Begin by introducing the idea that the parts of a boxplot divide data into quarters, with 25% of the values contained within each section. The three possible approaches are detailed below.

Using ICT

If students have access to a computer, set up a spreadsheet so that the cells are in a square format and an axis from 1 to 20 is visible (see Figure 12.2). Each cell represents a single data point. Ask students to colour cells so that for 60 data points the distribution of the data is equivalent to the boxplots on the worksheet. To help highlight the location of the five-number summary, they should use a different colour for each adjacent section.

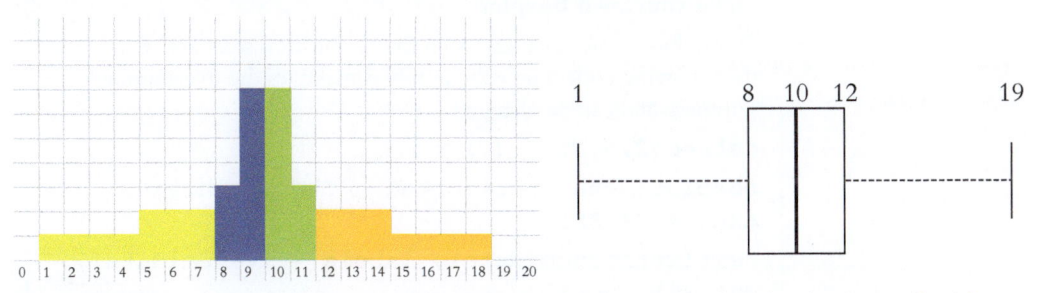

Figure 12.2 The graph on the left shows a possible distribution of 40 data points with the boxplot given on the right

Using paper

As with the Excel approach detailed above, provide students with squared paper; depending on the group you may wish to provide paper with the axis pre-drawn. Ask students to shade the squares with pencil/pen for the 60 data points. They should either use different colours for each adjacent section, or mark the key values for the five-number summary on their graph.

Using multilink cubes

Provide students with a sheet of plain paper on which a scale has been drawn from 0 to 20 with a unit the width of a multilink cube (or equivalent). Provide students with 60 cubes in four distinct colours (15 of each). Ask students to create the distribution using the cubes lined up on the axis. You may prefer to use a smaller number of cubes depending on availability, but the more cubes used, the smoother students can make the distribution.

Whichever approach is taken, for each boxplot students share their representations of the distributions and look for similarities and differences.

> **Discussion questions**
> - What common features do the graphs share?
> - Can you describe a scenario that fits your data?
> - Can you describe the variability of the data?

Additional tasks

Ask students to sketch a smooth curve to represent the graphs they have drawn and label the key values from the five-number summary.

Ask students to create a frequency table for the raw data implied by the distributions they have produced.

> **Make your own boxplots**
>
> To produce more boxplots using the R statistical software, use the following code. The italic numbers make up the five-number summary and can be changed.
>
> ```
> data=c(2,5,8,11,17)
> boxplot(data, horizontal = TRUE, range=0,
> axes = FALSE)
> text(x=fivenum(data), labels =fivenum(data),
> y=1.25)
> ```

Distribution worksheet

A boxplot is a diagram summarising a set of data. Each section of the boxplot contains 25% of the data points collected.

If 60 pieces of data are collected, there will be 15 individual pieces of data in each of the four sections.

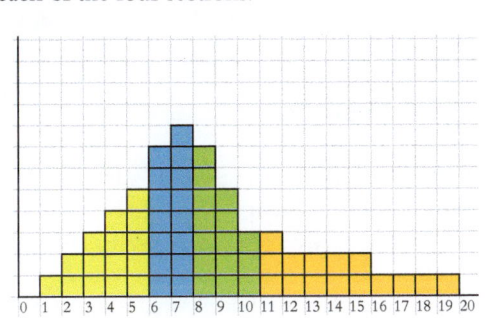

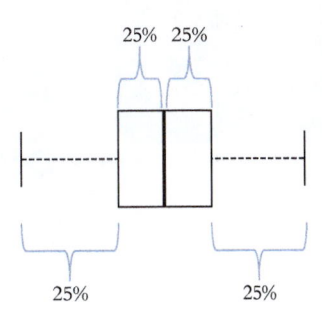

Each square on the graph represents one piece of data.
There are 60 pieces of data in total.

Each colour represents $\frac{1}{4}$ of the data points.

For each of the boxplots below, use the scale provided to decide where the 60 pieces of data could be located and create a graph using one square per piece of data.

Try to imagine a real-life scenario which could result in data that fits each of your graphs.

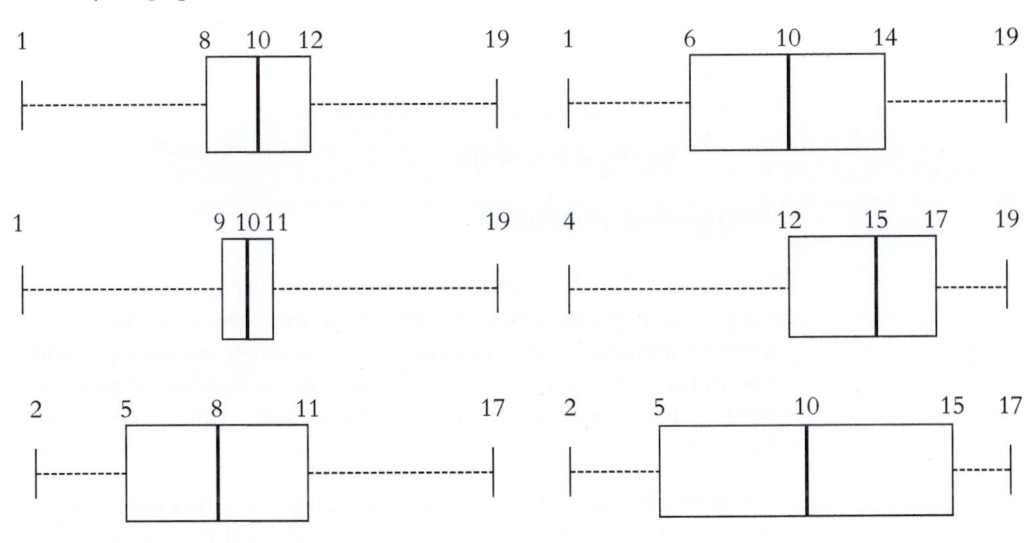

Chapter 13

Activity for Chapter 6: Signals and noise

Shoe size

Planning for the classroom

This activity is suitable for all levels, though more advanced students will have more techniques available to them and should be able to discuss in more detail. Shoe size is a relatively safe and easy-to-collect form of personal data that leads students to form hypotheses and draw conclusions about their own class. If your class is relatively small, say fewer than 20, it would be a good idea to collect data from the whole year group or school, or extend to family members. You can easily get a (minimally) multivariate data set by asking for other characteristics as well. For older or more able students, the extra fields raise some interesting questions about data representation. For the sake of simplicity we will keep this activity to just shoe size.

The whole idea of this task is to keep it technically very light touch. This is a guided discovery activity, setting the scene for formal calculation of summary data, estimators and the like later on.

Equipment and resources

Each student will need:

- a copy of the *Shoe size* worksheet
- data for class shoe size.

Suggested approach

This activity adds detail to the outline given in Chapter 6. Students collect a small personal data set of their shoe sizes, which are then sampled. Individual students receive their own sample and attempt to do two things: represent their data graphically; and find within the noise of their sample some conclusion they can draw about a typical shoe size for the class.

It is useful to have a particular aim for students so that they don't just wallow in the data. A good starter is 'What is a typical shoe size for a student in this class?' This will tend to focus them on case values first, then extending to attempts to define some sort of representative value, and hopefully, possibly with some prompts, to develop the idea of a measure of spread.

Stage 1 is to think about the class and how to find a representation of a typical shoe size, or of the shoe sizes of the group. Stage 2 attempts to generalise by asking 'What is a typical shoe size for a student in year X?', so that all the thinking done in terms of population is now about a sample and how we can use the parameters, or representations, of the sample to estimate, predict or represent the wider population.

Stage 1

Introduce the activity to the group with the simple question 'What is a typical shoe size for a student in this class?' Case value responses are very likely at this point: 'My shoes are size 5!', 'Bob has got tiny pixie feet!' and so on. This can quickly lead into further discussion.

> **Discussion questions**
> - Are these answers quantitative or qualitative?
> - How much information would this give an outsider about shoe size in the class?
> - What could we measure?
> - How can we record it?
> - How can we present it?
> - How do we separate the signal (typical shoe size) out from the noise (students calling out unconnected statements)?

There are essentially two options for measuring: measured shoe length or shoe size. Either will do perfectly well, and it is good to simply take the first offer from the class rather than correct it.

Aside: Another possibility, though unlikely to be offered except by older students, is to simply make pairwise comparisons and rank the students. Picking out the median student gives one possible signal, e.g. 'Susie represents the typical student in terms of shoe size', and you can create interesting visualisations using the students once they are all lined up in order by turning them into a human graph, photographing them as a group and projecting this onto a whiteboard if you have access to the technology.

The most likely approach, however, is that shoe length or shoe size is accepted and the class rapidly collects the data, giving you a small but very personal data set, in which each point represents a particular person.

The students now need some time to create a visual representation of the data, either individually or in groups. Give them the question again. The representation ought to attempt to answer this somehow, even if it is simply by annotating a bar graph. Try not to prompt particular traditional modes of representation; of course your students will use the various graph types encountered so far in the curriculum, but there are many creative alternatives. If left to their own devices, students may well add in other data fields such as gender or even physical position in the classroom.

Students should then pair up and compare representations. When critiquing each other's graph, the following questions could be used to prompt discussion.

> **Discussion questions**
> - How does the representation answer the question about the typical shoe size?
> - Does signal or noise dominate?
> - Is it still possible to pick out individual data points? Does this enhance the signal or create more noise?
> - Is the variation as noticeable as the central tendency?
> - What shape is the distribution? Is it skewed or symmetric? Does it have a tail? Does it have more than one mode?

Students might then like to improve or redo their graphs. Try to allow time for this. In a setting with relatively short lessons, this could be a good activity to take away as preparation for the next lesson.

Stage 2

Now for the generalisation steps.

Ask the class to imagine trying to answer the question 'What is a typical shoe size for a student in this class?' for someone outside the classroom without access to the data. Give them the task of answering this in three different ways by summarising and condensing the information.

1. By writing down a single number to communicate the typical shoe size.

2. By writing a full paragraph, using as many numbers and words as you need to.

3. By composing a 140 character tweet to communicate all the information.

Again, students should consider this individually or in pairs and then have time to discuss and critique. The purpose is to get students to think about what is needed to summarise the data fully by making them engage with lesser or greater restrictions on what they can report. Many students are familiar with attempting to condense information into a 140 character tweet, so this context provides a meaningful limitation. It may also lead to opportunities in future lessons to regularly show tweets from real organisations sharing data, to promote discussions that support development of statistical literacy.

Finally, we take a step through the looking glass. Having worked with the complete population data for their class, subtly switch the question to 'What is a typical shoe size for a student in your year group?' This would make a good plenary for overall discussion.

> **Discussion questions**
> - How can we use the data we have to create a graphical representation of shoe size for the whole year? How appropriate is it now for individual data points to be visible?
> - If you gave a single value as a typical representative value, to what extent can/does this represent the whole year group? What does the variation in the sample tell us about the accuracy of this value as an estimator?
> - If you gave a range, or a measure of spread, how does this measure of variation in the sample convert to that in the population?

Shoe size worksheet

In this activity you are going to answer the question 'What is a typical shoe size for a student in this class?'

Data presentation

Using the data for your class, produce a graphical representation which helps to answer the question 'What is a typical shoe size for a student in this class?'

Consider the following.

- How would your representation help someone to understand what a typical shoe size is?
- Is a graph on its own enough? Do you need to add some extra information?

Comparison and critique

Swap your graph with someone else and compare what you have done.

- How does *their* representation answer the question about the typical shoe size?
- Does the graph have any features that stand out?
- Is it still possible to pick out individual data points? Does this make it easier or harder to spot features of the whole data set?
- What changes would you make to your *own* graph as a result of looking at your partner's graph?

Summarising

By considering your graphical representation, try to answer the question 'What is a typical shoe size for a student in this class?' in the following three ways. Imagine that the reader can't see your graph.

1. By writing down a single number to communicate the typical shoe size.

2. By writing a full paragraph, using as many numbers and words as you need to.

3. By composing a 140 character tweet to communicate all the information.

Chapter 14

Activity for Chapter 7: Informal inference

The hidden world

Planning for the classroom

This activity is suitable for all levels. The aim is to help students develop their ability to use data and proportional reasoning to make inferences about a hidden population and then create a model based on their prediction. The activity should help students appreciate the link between proportion and probability from a statistical perspective. They will also experience the effects of changing sample size on stability. Because there are a lot of things going on in this activity, despite its simple structure, it is important to allow plenty of time for guided discussion in order to dig into the underlying mathematics.

Equipment and resources

You will need a spinner to conduct the experiment with. This consists of a printed copy of a circle split into different coloured sectors, a pen and a paper clip. *The hidden world resource sheet* provided at the end of this chapter gives a coloured and a blank template you could use. To construct the spinner, pull out one end of the paper clip, then pin the loop to the centre of the circle with a pen (see Figure 14.1). Flick the end of the paperclip and read off the colour on which the straightened-out end rests.

Figure 14.1 Construction of a simple spinner

Because this is a teacher-led activity, while you could produce printed prompt questions with spaces for answers, this is not necessary and the focus of the lesson should be on extensive student discussion rather than filling in a worksheet. *The hidden world worksheet* at the end of this chapter is a page of template spinners which can be given to students at the start of the activity.

139

Suggested approach

There are many alternative approaches to this activity depending on the preferred equipment. A possible approach is to use a bag containing coloured marbles or multilink cubes instead of a spinner.

Distribute the worksheet *The hidden world* to students in advance of the activity. Tell them that you want them to predict how the colours on the spinner are distributed without seeing the actual spinner – pose the question 'How could you do this?'

Guide the discussion towards the idea that you will spin the spinner and read off the colour that comes up, while students collect the data. Discuss how many spins they think would be needed to be sure what the colours were.

There is no need to decide as a class how many times the spinner should be spun; you will continue until everyone has a prediction they are happy with, and the purpose of the question is to prompt students to consider their initial assumptions. The worksheet has multiple blank spinners that they can use to make predictions by shading in the sections. Encourage students to make a prediction whenever they feel confident; they can then change their mind and make a new one if they change their view. You may want to pause every 10 spins and allow students the opportunity to make a prediction if they want to, and then either stick with or alter their prediction each time you pause.

Once all students have designed a possible spinner they are happy with, guide a discussion about how they could assess whether their predicted spinner is likely to be correct without actually seeing the original spinner. The aim here is for students to suggest repeating the experiment themselves with their own spinner and then compare their results to the original class experiment.

Allow students time to conduct their spinner experiment. Again there is no need to establish how many trials should be conducted; students can make this decision themselves but should be encouraged to record why they did so.

Once students have completed their experiments, discuss the individual predictions and the confidence students have in them after testing their models.

> **Discussion questions**
> - Do you still think your prediction is correct?
> - What evidence do you have that you were correct/incorrect?
> - How confident are you in your prediction? Why?

PART 2 Activities

Following the discussion, reveal the original spinner and discuss how closely it matches the predictions of the individual students. Focus on unusual results – why did students make predictions that turned out to be incorrect? Is there any link between people who made predictions based on a small number of trials? How did the number of trials affect their confidence in a prediction during the individual trials?

By the end of the activity students should have an increased awareness of the following points.

1   The proportion of each colour in the data becomes more stable as the number of trials increases.

2   A small number of trials (a small sample) is not a good basis for a prediction.

3   For a large number of trials, the proportion of each colour in the data reflects the proportion of each colour on the spinner.

4   In the real world of experimentation, sometimes it is impossible to know the underlying reality, but a model can be created from data and then tested. This is the basis of most scientific theory.

Appendix 1 explains how to create pie charts in the software package R, which is a quick and easy way of producing lots of spinners.

The hidden world resource sheet

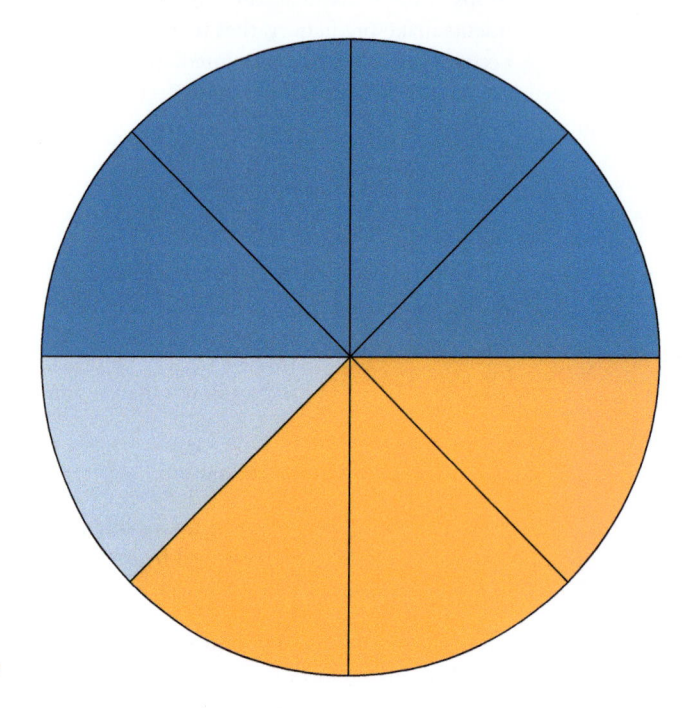

Basic spinner

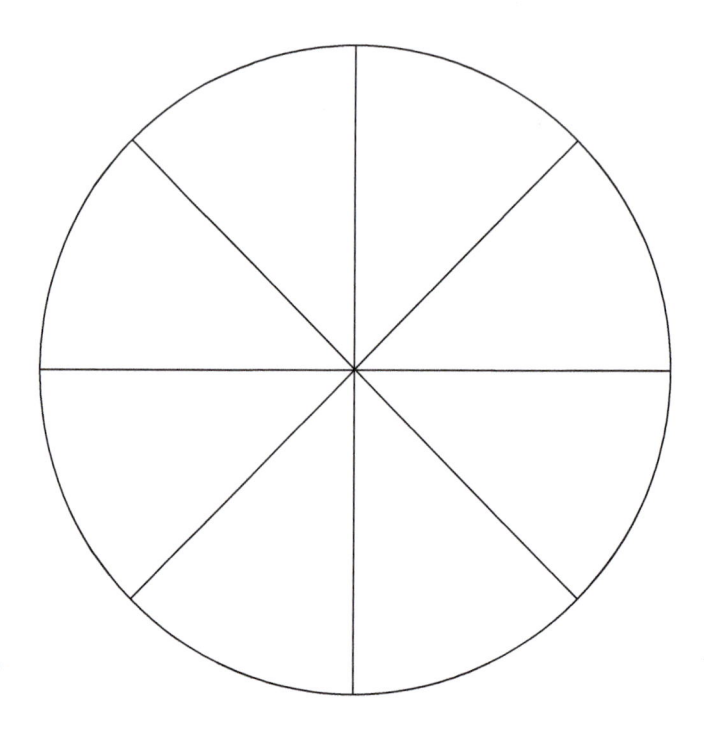

Blank spinner

PART 2 Activities

The hidden world worksheet

Predict the colours on the hidden spinner and fill in the blank spinners below. There are multiple blank spinners so that you can change your prediction at any time.

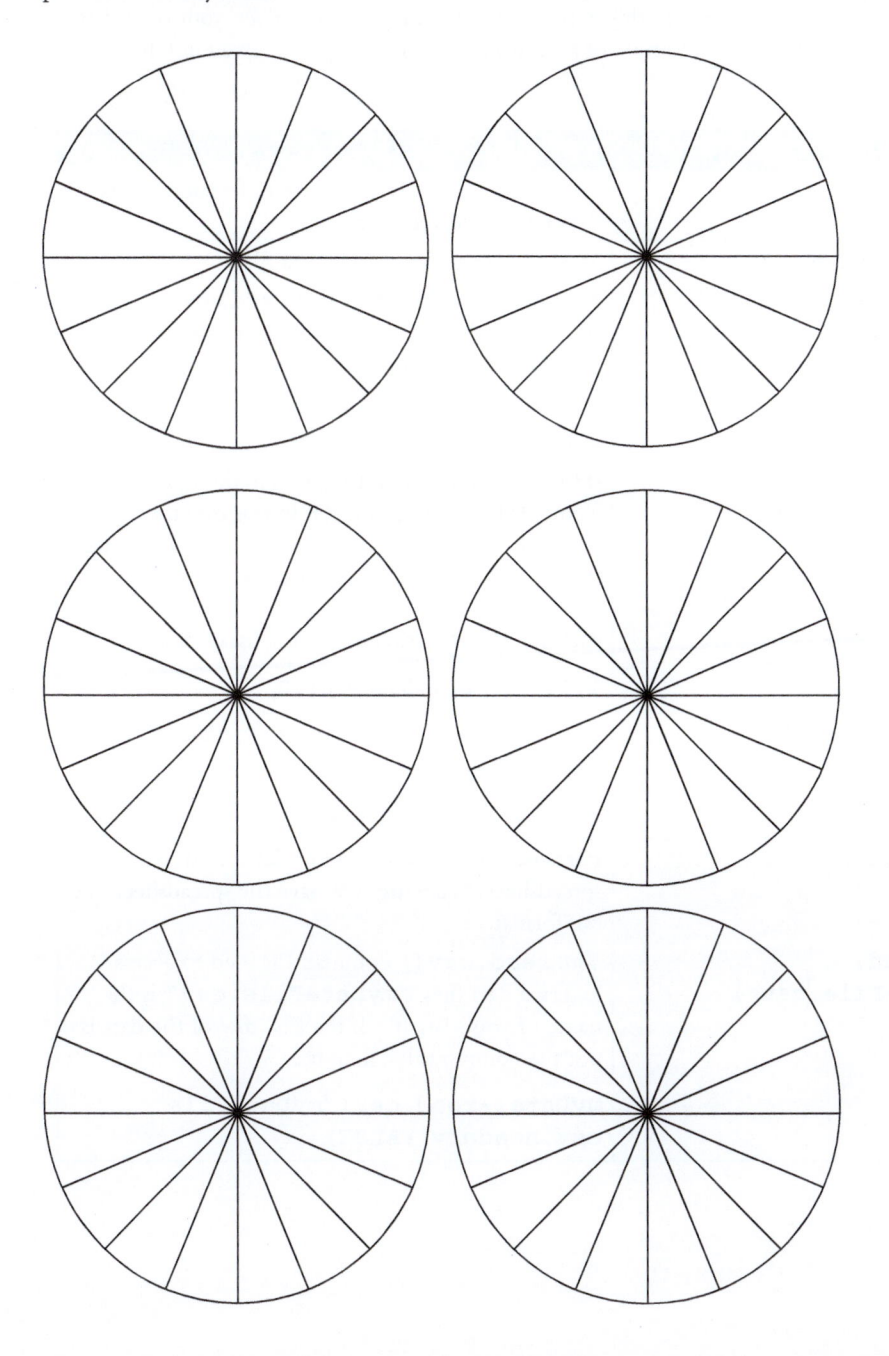

Appendix 1

Useful R commands

In the examples, we have used 'myData' as a generic name for a data set; you can replace this with whatever name you have given your own data. A detailed example of a classroom activity using R is given in Chapter 8.

Getting started	
Identify working directory `getwd()`	To find the directory where data files such as .csv data sets are currently stored.
Change working directory `setwd("/Users/yourname/statsfiles")`	To set the location of your data files. `"/Users/yourname/statsfiles"` is the location of the folder which contains your saved data. Must be in quotation marks.
Add data directly into R `myData<-c(data1,data2,data3,…)`	For when you have a short list of data that you want to enter by hand into the R console. `c()` is a function used to combine multiple elements into a single object. In this case the elements of a list are combined and assigned, using the `<-` operator, to a data object called 'myData'. `data1,data2,data3,…` is a list of data, e.g. `1,3,3,5,4,7,5,6,9,1,2` `"cat","dog","dog","cat","budgie", "cat","parrot"` Text data must be in quotes.
Add data from a .csv file `myData<-read.csv("myDataFile.csv")`	For when you want to import data saved in a spreadsheet. Make sure you save the spreadsheet in .csv format. The `read.csv()` command is used to access external data files. `"myDataFile.csv"` is the name of your data file. If the first row of the data is not the column titles, then use `MyData<-read.csv("myDataFile.csv,header= FALSE)`

Appendix 1

Getting started	
View your data `myData`	To show the data you have entered into R. Type the name you have given to your data object and hit enter.

View a single column of data from a table
 	
`myData$columnName`
	Type the name of the data object, followed by a `$` symbol and then the name of the column, e.g.
	`myData$height`
	`myData$frequency`
	`myData$column1`

Add new data points
 	
For univariate data:
`myData<-`
`c(myData,new1,new2,new3,…)`
For multivariate data:
`myData<-rbind(myData,c(new1a,new1b),c(new2a,new2b))`
	If you have an existing data set and want to add new values.
	
	`new1,new2,new3,…` are the new data to be entered.
	
	`rbind()` adds data as a new row; each comma within the parenthesis begins a new row of data. In this case, `c(new1a,new1b),…` are the new bivariate data added as new rows to myData.

Create a table from raw data `table(myData)`	If you want to plot a frequency plot but have entered your raw data as a list.
Drawing charts	
Create a bar chart `barplot(table(myData))`	[bar chart image] If your data is already entered as a table, use `barplot(mydata)`.

(continued)

145

Drawing charts	
Adding titles and labels to graphs	
The following parameters can all be used inside the parentheses of any `plot()` style command:	
`main=`	Adds a title to the graph.
`xlab=`	Adds a title to the *x*-axis.
`ylab=`	Adds a title to the *y*-axis.
	e.g.
	`barplot(table(myData),main="Number of brothers or sisters", xlab = " Number of siblings", ylab = "Frequency")`
Create a stacked dotplot	
`stripchart(myData, method = "stack", pch=1, offset = 0.5, at=0, frame = FALSE, xaxt="n")`	`method = "stack"` arranges the data vertically rather than overlaying repeat items.
	`pch=1` sets the symbol to a circle; try different numbers for different symbols.
	`offset = 0.5` sets the separation between the symbols; you can make it larger to spread the diagram vertically.
	`at=0` displays the lowest symbols close to the *x*-axis.
	The following parameters can be used in any plot:
	`frame = FALSE` removes the frame around the plot; this is not necessary but often desirable.
	`xaxt="n"` removes the standard *x*-axis, which is often not useful. See below for how to edit the axes directly.

(continued)

Appendix 1

Drawing charts	
Edit the axes `axis(1, at = seq(0,10, by = 1))`	To manually define an axis. 1 indicates the *x*-axis; use 2 for the *y*-axis. `at = seq(0,10, by =1)` creates an axis from 0 to 10, increasing by increments of 1.
Create a scatterplot `plot(myData)`	*[scatterplot of weight vs height]* Use `plot(myData, pch=4)` for crosses.
Add a best fit line `abline(lm(myData$yData ~ myData$xData))`	*[scatterplot with best fit line]* Type this command in after you have first created your scatterplot. `abline()` is the command that draws the graph. `lm()` instructs R to use a linear model. `xData` and `yData` are the names of the columns that hold your *x* and *y* data in the object mydata. So `myData$xData` is the *x*-variable of myData.
Create a boxplot `boxplot(myData, horizontal = TRUE)`	*[horizontal boxplot]* `horizontal=TRUE` prints the boxplot in the standard orientation. `boxplot(myData)` will print a vertical boxplot.

(continued)

147

Drawing charts	
Create a histogram `hist(myData)`	*[histogram plot]* To change the number of bars, use the parameter `breaks = n` e.g. `hist(myData, breaks = 10)` for 10 evenly split bars. For uneven classes, use `breaks` to tell R where the class boundaries are, e.g. `hist(myData,breaks =c(17,24,25,26,29,34))` `hist()` will decide automatically to use frequency density if classes are uneven.

Create a (ordered) case value plot
 	
Either
 	
`plot(sort(myData), type = "h")`
 	
or
 	
`barplot(sort(myData))`
	[case value plot]
	
	For lines (as above), if you leave out `type="h"` R will plot the data as points.
	
	If you prefer bars, use `barplot`.
	
	`sort(mydata)` orders the data.
	
	Just use `myData` for unordered data.

(continued)

Appendix 1

Drawing charts	
Create a pie chart `pie(table(myData))`	If your data is already in table format use `pie(myData)`

Create a time-series graph
 	
These instructions assume that the data object was imported from a 2 column csv file with column 1 a list of times, e.g. months, and column 2 the variable to plot.
 	
`myTsData<-`
`ts(myData$plotVariable)`
	`myTsData` is a new time-series data object which can take any name. The `ts()` command identifies the data as a time-series.
	
	`myData$plotVariable` is the column of measurements being plotted, e.g. `myData$rainfall`, `myData$iceCreamSales`

`plot(myTsData, xaxt = "n")`	`plot()` creates the graph of the time-series object. `xaxt= "n"` suppresses the default axis.
`axis(1, at = 1:n, labels = myData$timeVariable)`	`axis()` creates a custom axis. `1` is the key for the *x*-axis. `at=1:n` identifies the location of the tick marks on the axis, replace `n` with the number of variables being plotted, e.g. `at = 1:12` for a full year in months. `labels = myData$timeVariable` assigns the values in the time column of your original data object to the tick marks on the *x*-axis.

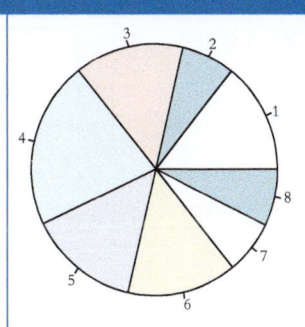

(continued)

Drawing charts

Create a cumulative frequency graph	*(cumulative frequency graph showing values 0–250 on y-axis against mybreaks 2–5 on x-axis)*
Write each line of code and then press enter:	This works for raw, ungrouped data.
`myBreaks = seq(lower,upper, by = step)`	`myBreaks` can be any name you want to give to the boundaries of each data class.
	The `seq()` command generates a list of numbers where lower is the smallest number, upper is the largest number and steps is the increment, e.g. `seq(1, 5, 0.5)` generates 1, 1.5, 2, 2.5, 3, 3.5, 4, 4.5, 5
`myData.cut = cut(myData, myBreaks, right = FALSE)`	`myData.cut` is the name of a new data object and can be any word you choose. The `cut()` command takes each piece of data in myData and assigns it to one of the classes defined by the class boundaries in myBreaks, e.g. a value 1.27 is assigned to the group $1 < x \leq 1.5$.
	`right = FALSE` defines the upper bound of each group as $\leq$
`myData.freq = table(myData.cut)`	`myData.freq` is the name of a new data object and can be any word you choose.
	`table()` creates a frequency table counting the instances of each class created in `myData.cut`.

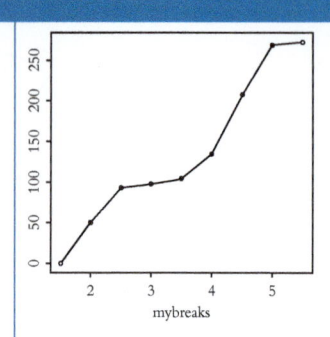

(continued)

Appendix 1

Drawing charts	
`myCumFreq = c(0,cumsum(myData.freq))`	`myCumFreq` is the name of a new data object and can be any word you choose. The `c()` command combines the following: `0` adds a zero value to the lower end of the data; `cumsum(myData.freq)` converts the frequency table to a cumulative frequency table.
`plot(myBreaks, myCumFreq)`	Plots the points of the cumulative frequency against the boundaries in `myBreaks`.
`lines(myBreaks, myCumFreq)`	Connects the points with a straight line.
Create a frequency polygon Write each line of code and then press enter:	

`myBreaks = seq(lower,upper, by = step)`
`myData.cut = cut(myData, myBreaks, right = FALSE)`
`myData.freq = table(myData.cut)`
`myFreqPoly = c(0, myData.freq)`
`plot(myBreaks, myFreqPoly)`
`lines(myBreaks, myFreqPoly)`
	This works in exactly the same way as for a cumulative frequency graph; however, instead of `myCumFreq`, create `myFreqPoly` without the `cumsum()` command.

(continued)

151

Drawing charts	
Create a scatterplot matrix	For creating several scatterplots of paired variables in a multivariate data set.
`pairs(~col1+col2+col3+col4, myData)`	The `pairs()` command creates a grid of scatter graphs for each column listed inside the parentheses. `~col1+col2+`... are the names of the columns to be included, `~` and `+` are necessary syntax. `myData` is the data source.
Calculating statistics	
Select a random sample from a population `sample(myData,samplesize)`	For selecting a small sample of data from a larger data set. `samplesize` can be any number, e.g. `sample(myData,10)` gives a random sample of 10 data points.
Calculate a five-number summary `summary(myData)`	For creating a five-number summary of the data: minimum value, lower quartile, median, upper quartile, maximum value.
Calculate mean `mean(myData)`	To calculate the arithmetic mean.
Calculate standard deviation `sd(myData)`	To calculate the standard deviation.
Calculate correlation coefficient `cor(myData$column1, myData$column2)`	To calculate a correlation coefficient. `column1` and `column2` are the column headings from your data.

(continued)

Appendix 1

This gives just a brief flavour of some of the commands in R that may be useful for quickly presenting data to students in statistics lessons. Expert users of R may look at many of the commands here and criticise them as sub-optimal ways of working with R; we have attempted to keep the commands as simple as possible at the expense of some optimisation in order to make them accessible to those unfamiliar with this type of software and interface.

For more detailed information, advanced features and data objects, there are many online guides and communities that can help.

Useful links

https://cran.r-project.org/manuals.html

https://stat.ethz.ch/R-manual/

https://www.r-bloggers.com/

https://little-books-of-r.readthedocs.io/en/latest/#

Appendix 2

Some prompts for investigations

Many examination systems have dabbled with the idea of an investigative component to the terminal assessment. This appendix contains a selection of statistical investigation prompts that made up one component of a mathematics GCSE examination provided by a UK-based awarding organisation. These kinds of activities are an excellent source of data and ideas which provide opportunities for students to explore data and contexts.

At the beginning of each task, students were provided with the following information.

Guidance for candidates

To gain the highest mark possible, you should attempt to include as many of the following features as possible.

Plan the task
- State clearly what your aims are before you work through the task.
- Write a plan that allows you to find out as much as possible about the task.
- State where you will obtain your data.
- If you have to sample data, say how and why you chose the sample, how this might affect your results and what you may have to do to overcome any problems.
- Use correct statistical terms at all times.

Explain your work
- Show clearly what you have done with your data.
- When you use calculations, show the working you had to do.
- If you use ICT, include print-outs and explain clearly what the graphs, tables and any figures calculated tell you about your work.
- Use only calculations, graphs or tables that are useful.
- Check carefully that you have not made errors.

State your findings	• Try to write your comments near to the calculations, graphs or tables that you have produced.
	• Write comments that explain what your results tell you about the task.
	• Say how effective your plan was in helping you find out about the task.
	• Say what realistic improvements you could have made to improve the method(s) you used.

For each task students were provided with a scenario, a handful of prompt questions and a set of data, along with a suggestion to continue or extend the investigation. When using these tasks in the classroom, you may choose to provide more or less structure, depending on the ability and experience of the students. You may also encourage students to collect their own data rather than using the data provided.

Task 1: How good are your reactions?

In some countries traffic lights change straight from red to green. The time a motorist takes to react to this change is their **reaction time**.

Experiments have been carried out to measure reaction times. One such experiment involves measuring the time taken to press a button when a light changes from red to green.

Write at least one hypothesis about people's reaction times. Use appropriate data and techniques to test your hypotheses, planning and specifying your methods carefully and stating your aims clearly.

- A sample of data for the experiment described above is given in the following table.

Appendix 2

Data

Reaction times in seconds.

All data from UK Phase 4 CensusAtSchool.

NB This is raw data and may well contain some errors!

Region	Gender	Age in years	Hand normally used for writing	Reaction, left hand (seconds)	Reaction, right hand (seconds)
East Midlands	F	15	left	0.22	0.22
East Midlands	F	15	right	0.22	0.22
East Midlands	M	15	right	1.7185	0.2575
East Midlands	M	15	right	0.385	0.36
East Midlands	M	15	right	0.19	0.11
East Midlands	F	15	right	0.22	0.22
East Midlands	F	15	right	0.3	0.22
East Midlands	F	15	right	0.3	0.305
East Midlands	M	15	right	0.33	0.355
East Midlands	F	15	right	0.305	0.305
East Midlands	M	15	right	0.2265	0.2735
East Midlands	M	15	right	0.3	0.16
East Midlands	M	15	right	0.22	0.245
East Midlands	M	15	right	0.2145	0.19
East Midlands	F	15	right	0.88	0.355
East Midlands	F	15	right	0.385	0.3
East Midlands	F	15	right	0.28	0.305
East Midlands	F	15	right	0.28	0.22
East Midlands	M	15	right	0.22	0.135
East Midlands	M	15	left	0.22	0.19
East Midlands	F	15	right	0.335	0.385
East Midlands	M	15	right	0.25	0.275
East Midlands	F	15	right	0.305	0.415

Region	Gender	Age in years	Hand normally used for writing	Reaction, left hand (seconds)	Reaction, right hand (seconds)
East Midlands	F	15	right	0.266	0.241
Home Counties	F	15	right	0.28	0.275
London	F	15	right	0.245	0.22
North East	F	15	left	0.133	0.3595
North West	F	15	right	0.25	0.22
North West	M	15	left	0.1955	0.111
North West	M	15	right	0.6005	2.6585
North West	M	15	right	0.6005	2.6585
North West	F	15	right	0.2305	0.24
North West	M	15	right	0.41	0.25
North West	F	15	right	0.4925	0.2655
North West	M	15	right	0.164	0.047
North West	F	15	right	0.321	0.266
North West	F	15	right	0.25	0.203
North West	F	15	right	0.24	0.33
North West	M	15	right	0.25	0.21
North West	M	15	right	0.225	0.245
North West	M	15	right	0.255	0.255
North West	M	15	right	0.6355	0.3105
Scotland	F	15	right	0.3	0.245
South	F	15	left	0.19	0.21
South	M	15	right	0.22	0.2055
South	F	15	left	0.245	0.2005
South	F	15	right	0.2755	0.19
South	M	15	right	0.245	0.22
South	M	15	right	0.2605	1.062
South	M	15	right	0.2155	0.1655
South East	F	15	right	4.745	3.98

(continued)

Appendix 2

Region	Gender	Age in years	Hand normally used for writing	Reaction, left hand (seconds)	Reaction, right hand (seconds)
South Wales	F	15	right	0.275	0.245
West Midlands	F	15	right	0.211	0.2185
West Midlands	M	15	right	0.188	0.1875
West Midlands	F	15	right	0.2265	0.211
West Midlands	F	15	right	0.203	0.203
West Midlands	F	15	right	0.203	0.1015
West Midlands	M	15	left	0.2265	0.203
East	F	12	right	0.22	0.22
East	F	12	right	0.435	0.335
East	M	12	right	0.11	0.115
East	F	12	right	0.22	0.215
East	M	12	left	0.55	0.465
East	F	12	left	0.33	0.3
East	F	12	right	0.77	0.38
East	F	12	left	0.165	0.22
East	F	12	right	0.601	0.29
East	F	12	left	0.305	0.275
East	F	12	either	2.665	0.74
East	F	12	right	0.195	0.28
East Midlands	F	12	right	0.3205	0.326
East Midlands	F	12	right	0.2755	0.285
East Midlands	M	12	left	0.2425	0.4295
East Midlands	F	12	right	0.355	0.25
East Midlands	F	12	right	0.242	0.508
East Midlands	M	12	either	0.305	0.96
East Midlands	M	12	right	0.465	0.27
East Midlands	F	12	left	0.265	0.306
East Midlands	M	12	right	0.71	13.21

(continued)

Region	Gender	Age in years	Hand normally used for writing	Reaction, left hand (seconds)	Reaction, right hand (seconds)
East Midlands	M	12	right	3.0015	4.9195
East Midlands	F	12	right	0.331	0.3365
East Midlands	F	12	right	0.25	0.25
East Midlands	F	12	right	0.33	0.36
East Midlands	F	12	right	0.221	0.265
East Midlands	M	12	left	1.345	0.715
East Midlands	M	12	right	0.235	0.2305
East Midlands	M	12	right	0.985	1.925
East Midlands	F	12	right	0.2555	0.2605
East Midlands	M	12	right	0.256	0.311
East Midlands	M	12	right	0.5395	1.7815
East Midlands	F	12	right	0.651	0.746
East Midlands	M	12	left	2.6085	1.046
East Midlands	F	12	right	1.142	0.2655
East Midlands	M	12	right	0.245	0.271
East Midlands	M	12	right	0.1015	1.828
East Midlands	M	12	right	0.2605	0.281
East Midlands	F	12	right	0.32	0.4145
East Midlands	F	12	right	2.6435	0.325
East Midlands	M	12	right	0.33	0.245
East Midlands	M	12	right	0.3	0.495
East Midlands	M	12	right	0.36	0.41
East Midlands	M	12	right	0.175	0.236
East Midlands	F	12	right	0.275	0.245
East Midlands	M	12	right	0.328	0.328
East Midlands	F	12	left	0.1605	0.5155
East Midlands	F	12	either	0.276	0.266
East Midlands	M	12	right	0.195	0.21

(continued)

Appendix 2

Region	Gender	Age in years	Hand normally used for writing	Reaction, left hand (seconds)	Reaction, right hand (seconds)
East Midlands	F	12	right	0.25	0.2745
East Midlands	F	12	right	0.24	0.235
East Midlands	F	12	right	0.285	0.2405
East Midlands	F	12	right	0.291	0.536
East Midlands	F	12	left	0.3005	0.2905
East Midlands	F	12	left	0.2855	0.3005
East Midlands	F	12	right	0.563	0.2655
East Midlands	M	12	right	0.63	0.435
Home Counties	M	12	right	1.0715	0.892
Home Counties	M	12	right	0.25	0.211
Home Counties	M	12	right	0.495	0.25
Home Counties	F	12	right	2.5785	1.1715
Home Counties	M	12	right	0.2345	0.1955
Home Counties	F	12	right	0.2735	0.2575
Home Counties	M	12	right	0.215	0.025
Home Counties	M	12	right	0.41	0.245
Home Counties	F	12	right	0.275	0.275
Home Counties	F	12	right	0.242	0.289
Home Counties	M	12	left	0.28	0.245
Home Counties	F	12	right	0.325	0.385
Home Counties	M	12	right	0.2205	0.221

(continued)

[OCR, GCSE Mathematics, J512/06/B254/B266, 2007]

Mark scheme

Below is a copy of the marking frame which was used to mark all of the tasks in Appendix 2. The tasks were marked against three separate strands, assessing the student's ability to 'specify and plan', 'collect, process and represent', and 'interpret and discuss' by assigning a mark from 1 to 8 based on the quality and sophistication of the student's responses. The examples in the marking frame relate to Task 1, but the frame can be used to assist in marking all of the tasks by applying the general criteria listed in the first two columns.

It should be noted that the marking frame is provided in this appendix to support teachers in awarding credit to students for their investigative work but should not be treated as a straightjacket, used to confine students' responses to only those which can be predicted by the original assessors. A large dose of common sense must be applied along with the marking criteria to ensure that where students have creatively applied valid techniques and come up with unanticipated insights, they are not marked down for not having 'guessed' in advance what the mark scheme should be.

Similarly we do not recommend sharing the marking frame with students in advance of their work, to avoid it being seen as a 'tick list' of criteria to meet in order to gain maximum credit. They should be free to explore the data, select techniques and make the decisions that result, rather than being channelled down a narrow path towards a pre-imagined successful conclusion.

Specify and Plan

Notes:
1. In these criteria there is an intended approximate link between 7 marks and grade A, 5 marks and grade C and 3 marks and grade F.

2. Candidates must provide evidence of their plan being implemented.

3. If secondary data is provided it must be of sufficient quantity to allow sampling to take place.

		Minimum requirements	Examples
1	Candidates choose a simple well-defined problem. Their aims have some clarity. The appropriate data to collect are reasonably obvious. An overall plan is discernible and some attention is given to whether the plan will meet the aims. The structure of the report as a whole is loosely related to the aims.	• The candidate shows they understand a simple problem. • There is an implicit plan.	**Implicit** plan and some further work. Eg • May select some of the data and attempt to average. • May add up some of the data or rank it. • May draw a frequency diagram for some of the data.
2			As for S1 but gives some structure to the write up. May aim to compare two simple samples possibly by finding two averages or drawing bar charts.

Appendix 2

		Minimum requirements	Examples
3	Candidates choose a problem involving routine use of simple statistical techniques and set out reasonably clear aims. Consideration is given to the collection of data. Candidates describe an overall plan largely designed to meet the aims and structure the project report so that results relating to some of the aims are brought out. Where appropriate, they use a sample of adequate size.	• Candidates set out reasonably clear aims (or the purpose). • Their planning is largely designed to meet the aims/purpose. • They use data appropriate to the problem.	Writes a very brief outline plan and may intend calculation of the average reactions for two (or more) different groups. (Male, female or age 12 and age 15). States one simple aim. Eg • Compare the average (or spread) of the reaction times or • Tally the data into groups and draw a frequency diagram so that the most common time may be seen.
4			S3 and a clear structure to meet the stated aim. Indicates how the aim may be met through the way techniques will be used.
5	Candidates consider a more complex problem. They choose appropriate data to collect and state their aims in statistical terms with the selection of an appropriate plan. Their plan is designed to meet the aims and is well described. Candidates consider the practical problems of carrying out the survey or experiment. Where appropriate, they give reasons for choosing a particular sampling method. The project report is well structured so that the project can be seen as a whole.	• Candidates consider a **substantial** problem stating their initial aims clearly at the beginning of the report. • Their plan is explicitly stated to meet those aims. • They choose an appropriate sample.	States a clear plan with one or more aims in general terms (may involve design of an experiment). • Aims to compare two or more RELATED factors that may affect reaction times (age, hand, time, area,…). • Selects appropriate data to complete the task.
6			S5 and the plan is well structured, with some reasoning for the plan, and uses statistical terms to state each subtask's aims. • Eg Compare medians and IQRs for different genders with a view to determining more than a simple average comparison. The data is chosen with some thought to the avoidance of bias.

(continued)

		Minimum requirements	Examples
7	Candidates work on a problem requiring creative thinking and careful specification. They state their aims clearly in statistical terms and select and develop an appropriate plan to meet these aims giving reasons for their choice. They foresee and plan for practical problems in carrying out the survey or experiment.	• Candidates work on a **demanding** problem. • They state their aims clearly in statistical terms and give valid reasons for their choice of planning. • They explain and act upon limitations of their chosen sample (eg bias) where appropriate.	Chooses three or more RELATED subtasks that explore reaction times. (May involve design of an experiment) • Eg Gender, Age, Time of day (type of stimulus). The strategy is well planned and utilizes appropriate techniques and choice of data. Statistical terms are used to state each subtask's aims. Plans to draw the results together.
8	Where appropriate, they consider the nature and size of sample to be used and take steps to avoid bias. Where appropriate, they use techniques such as control groups, or pre-tests or questionnaires or data sheets, and refine these to enhance the project. The project report is well structured and the conclusions are related to the initial aims.		S7 and there is an efficient plan to achieve the aims in each subtask. These are all designed to explore one, overarching, hypothesis. Eg "People react differently to different stimuli", OR "People react differently at different times of day". Choices and plans are justified and statistical language is consistently and accurately used.

(continued)

Collect, Process and Represent

Notes: 1 In these criteria there is an intended approximate link between 7 marks and grade A, 5 marks and grade C and 3 marks and grade F.

   2 The mark awarded to a particular technique should reflect the quality of use and understanding as well as its position within the Level Indicators.

   3 The inclusion of statistical techniques outside the National Curriculum does not necessarily justify the award of higher marks.

4 'Diagrams' include tables, charts and graphs. At 5-6 marks the diagrams used should be appropriate. At 7-8 marks the range of diagrams should be appropriate to the problem chosen and the statistical strategy chosen.

5 'Redundancy' implies unnecessary and/or inappropriate diagrams or calculations. This includes techniques that are not used for any conclusion.

		Minimum requirements	Examples
1	Candidates collect data with limited relevance to the problem and plan. The data are collected or recorded with little thought given to processing. Candidates use calculations of the simplest kind. The results are frequently correct. Candidates present information and results in a clear and organised way. The data presentation is sometimes related to their overall plan.	• Candidates collect or use data and record it.	Shows some working towards achieving a mean or mode or range from some of the given data or some tally.
2			As C1 but with well-organised working or drawing of a tally table for one subset of the data, with a "correct" result. May include one (or two) bar chart(s).
3	Candidates collect data with some relevance to the problem and plan. The data are collected or recorded with some consideration given to efficient processing. Candidates use straightforward and largely relevant calculations involving techniques meeting the level detailed in the handling data paragraph of the grade description for grade F. The results are generally correct. Candidates show understanding of situations by describing them using statistical concepts, words and diagrams. They synthesise information presented in a variety of forms. Their writing explains and informs their use of diagrams, which are usually related to their overall plan. They present their diagrams correctly, with suitable scales and titles.	• Candidates collect or use data with some relevance to the problem. • They utilise statistical techniques/ diagrams (see note 1 above) to process and represent the data. • Their results are generally correct.	Carries out their plan, showing table(s) and calculation(s), finding one (or more) of mean, mode or range for at least one subset of the data. May represent the subset of data in a frequency diagram, including some comment(s) to indicate what has been done.
4			C3 and there is a clear linking commentary that **synthesises** their results.

165

		Minimum requirements	Examples
5	Candidates collect largely relevant and mainly reliable data. The data are collected in a form designed to ensure that they can be used. Candidates use a range of more demanding, largely relevant calculations that include techniques meeting the level detailed in the handling data paragraph of the grade description for grade C. The results are generally correct and no obviously relevant calculation is omitted. There is little redundancy in calculation or presentation. Candidates convey statistical meaning through precise and consistent use of statistical concepts that is sustained throughout the work. They use appropriate diagrams for representing data and give a reason for their choice of presentation, explaining features they have selected.	• Candidates collect/ sample largely relevant data. • They utilise appropriate calculations/ diagrams/ techniques (see note 1 above) within the problem. • Their results are generally correct.	Uses appropriate techniques linked to their S4 (at least) plan. These are likely to include mean and range for at least two subsets of data (given and collected). At least 30 reaction times considered. Data may be grouped, estimated mean may be calculated, comparative frequency diagrams or scatter charts may be used, spreadsheets may be used to perform calculations and generate appropriate graphs.
6			As C5 but may include ogive, IQ range, box and whisker plots and these are used consistently and appropriately. Clear understanding is shown and there is some justification for the choice of diagrams and calculations.

(continued)

Appendix 2

		Minimum requirements	Examples
7	Candidates collect reliable data relevant to the problem under consideration. They deal with practical problems such as non-response, missing data or ensuring secondary data are appropriate. Candidates use a range of relevant calculations that include techniques meeting the level detailed in the handling data paragraph of the grade description for grade A. These calculations are correct and no obviously relevant calculation is omitted. Numerical results are rounded appropriately. There is no redundancy in calculation or presentation. Candidates use language and statistical concepts effectively in presenting a convincing reasoned argument. They use an appropriate range of diagrams to summarise the data and show how variables are related.	• Candidates collect/ sample largely relevant data. • They utilise appropriate and necessary calculations/ diagrams/ techniques (see note 1 above) consistently within the problem. • Their results are correct (some minor errors may be condoned provided they do not detract from the quality of the argument).	The candidate selects or gathers data that is reliable and relevant to the designated subtasks so that they may meet the aims of their S7 (or well-structured S6) hypothesis. TWO or more grade B techniques have been appropriately applied and the outcomes of these correctly interpreted, in the light of the problem. Presentation justified.
8			As C7 but with **efficient** and also correct use of a grade A technique and language to present an **argument**, in statistical terms, based upon the data analysis.

(*continued*)

167

Interpret and Discuss

Notes:
1. In these criteria there is an intended approximate link between 7 marks and grade A, 5 marks and grade C and 3 marks and grade F.

2. The number of marks awarded in this strand is unlikely to exceed the mark in strand 1 by more than 1 mark.

		Minimum requirements	Examples
1	Candidates comment on patterns in the data. They summarise the results they have obtained but make little attempt to relate the results to the initial problem.	• Candidates comment on their data.	Very limited comments such as "The slowest reaction time is…"
2			Makes a comment based upon their results, eg "The ten year olds were slower than the fifteen year olds."
3	Candidates comment on patterns in the data and any exceptions. They summarise and give a reasonably correct interpretation of their graphs and calculations. They attempt to relate the summarised data to the initial problem, though some conclusions may be incorrect or irrelevant. They make some attempt to evaluate their strategy.	• Candidates summarise some of their data. • They make a statement based on their diagrams or calculations, which is relevant to the problem.	Most likely linked to S3. May produce a table showing all the averages for two groups and writes a general comment related to these. Eg "My results show that the fifteen year olds have lower average reaction times than the ten year olds."
4			I3 and more specific statements that relate directly to the aims. Eg "The mean for the fifteen year olds was 0.1 seconds lower than the mean for the ten year olds. This shows that…"

Appendix 2

		Minimum requirements	Examples
5	Candidates comment on patterns in the data and suggest reasons for exceptions. They summarise and correctly interpret their graphs and calculations, relate the summarised data to the initial problem and draw appropriate inferences. Candidates use summary statistics to make relevant comparisons and show an informal appreciation that results may not be statistically significant. Where relevant, they allow for the nature of the sampling method in making inferences about the population. They evaluate the effectiveness of the overall strategy and make a simple assessment of limitations.	• Candidates summarise **and** correctly interpret their diagrams and calculations. • They relate these interpretations back to the original problem. • They evaluate their strategy.	Provides a clear interpretation for their calculations and diagrams. Makes simple evaluative statements that recognise strengths or weaknesses in their strategy. May clearly compare different groups through their means and ranges. May comment about the shape of grouped frequency (comparative) diagrams and link these to the means calculated.
6			I5 and makes statements that involve reference to measures calculated within the task that relate to their aims. Makes statements of evaluation and begins to give reasons for WHY these would improve their strategy and the outcomes of the work.

(continued)

		Minimum requirements	Examples
7	Candidates comment on patterns and give plausible reasons for exceptions. They correctly summarise and interpret graphs and calculations. They make correct and detailed inferences from the data concerning the original problem using the vocabulary of probability. Candidates appreciate the significance of results they obtain. Where relevant, they allow for the nature and size of the sample and any possible bias in making inferences about the population. They evaluate the effectiveness of the overall strategy and recognise limitations of the work done, making suggestions for improvement. They comment constructively on the practical consequences of the work.	• Candidates summarise and correctly interpret their results. • They show an appreciation of the significance of these results. • They recognise possible limitations in their strategy and suggest improvements.	**Most likely S7 but a good case of S6 may be considered.** Correct statements of interpretation of the findings from techniques applied to their subtasks. These subtasks are drawn together and are not a series of separate components. Some statements of evaluation, relating to improvements that could be made, are included and these are justified.
8			**S7 is expected but you may award on a good S6.** I7 and further analysis of the strategy. Suggests realistic improvements to the work and justifies these. Accounts for any bias in sampling. Sophisticated statements of interpretation and evaluation are made. Statistical language is used concisely to convey meaning.

(continued)

Task 2: SPELL

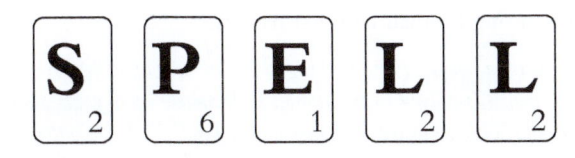

S₂ **P**₆ **E**₁ **L**₂ **L**₂

SPELL is a word game played using lettered cards.

Points are scored for making words using the cards.

Each letter has a points value.

These are the points values for each of the letters.

Letter	Points
A	3
B	2
C	2
D	1
E	1
F	1
G	1

Letter	Points
H	5
I	2
J	4
K	5
L	2
M	4
N	5

Letter	Points
O	3
P	6
Q	4
R	1
S	2
T	10
U	9

Letter	Points
V	2
W	9
X	5
Y	1
Z	6

The score for a complete word is the total of all the letter points.

For example, SPELL scores 13 points.

The task is about researching the appropriateness of the scores given to letters in the game of SPELL.

1. In the game of SPELL

   a. which letters of the alphabet are given the lowest values?

   b. which letters of the alphabet are given the highest values?

Letters that are easy to use should have a low value.

2. Have suitable values been given to **each** letter of the alphabet?

3. Extend your work on the task appropriately. Make clear your plan, reasons and conclusions.

[OCR, GCSE Mathematics, 1962/08/2345/2318, 2002]

Task 3: Sports report

Maria was discussing football with her friends.

'English football is so boring. For one thing, there are so few goals scored. It is more exciting when it is played in the Italian style, where more goals are scored. No wonder people don't go to watch games in England.'

Not everyone agreed.

Here are two sets of results from the period September 28 to 30 in the 2002–2003 season.

English Premier Division

Home	Scores	Away	Attendance	Time to first goal (mins)
Tottenham	0 - 3	Middlesbrough	36 082	32
Sunderland	1 - 0	Aston Villa	40 492	76
Man City	0 - 3	Liverpool	35 141	4
Leeds	1 - 4	Arsenal	40 199	20
Everton	2 - 0	Fulham	34 371	45
Chelsea	2 - 3	West Ham	38 929	21
Charlton	1 - 3	Man Utd	26 630	43
Bolton	1 - 1	Southampton	22 692	82
Birmingham	0 - 2	Newcastle	29 072	34
West Brom	0 - 2	Blackburn	25 170	72

Italy Series A

Home	Scores	Away	Attendance	Time to first goal (mins)
Bologna	1 - 0	Piacenza	22 000	66
Brescia	2 - 3	Roma	20 000	30
Como	1 - 1	Reggina	7 000	17
Modena	2 - 1	Torino	14 000	28
Perugia	1 - 3	Empoli	7 000	7
Udinese	1 - 0	Atalanta	15 000	59
Inter-Milan	2 - 1	Chievo	57 674	2
Juventus	2 - 2	Parma	40 000	66
Lazio	1 - 1	AC Milan	45 000	7

1 Look at the evidence in the tables. Is Maria right to think that more goals are scored in Italian football than in English football?

2 Extend your investigation of English football using other factors. Make clear what your hypothesis is, how you plan to research it, and your conclusions.

[OCR, GCSE Mathematics, 1962/08/1966/2345/1968/2318/1969/05, 2003]

Task 4: Estimate

Brian asked 120 people to estimate the area of the rectangle drawn opposite.

These are the estimates people gave to Brian for the area of the rectangle.

All measurements are in square centimetres.

90.5	140.6	180	151.55	112	129.7
100	134.6	168.3	201.3	146.4	132
189	152.25	126.3	141.3	165	147.4
170	188	136.2	144.8	139	126
41.2	100	129	156.2	146.8	146.9
141	111	134.8	155.5	183	157
157.7	161.5	144	119.1	125	125
130	132.2	147.9	151.6	159	154.2
122	129	139.5	146	132	133.7
146.6	155	146.8	158.9	165.3	119
143	121.25	115	166	183	108.2
155.8	149	148.2	146.3	175.9	157.3
118	170	138	129	148.5	166
169.3	137.1	142.6	134.2	93	167.4
111.3	147.3	160	141.8	177	100
139	177.4	184	140	146.2	131.8
135.5	135.5	145	175	250	122
166.2	138	135.8	160	151	133
95	145.4	190	162.6	140	125
147.8	149.5	164	137.4	121.2	169

The area of the rectangle is 145.36 cm².

1 Analyse the estimates that were made for the area of this rectangle.

2 Extend your investigation, making clear the rules and methods that you use.

Appendix 2

Can you estimate the area of this rectangle?

[OCR, GCSE Mathematics, 1962/08/2345/2318/1969/05, 2004]

Task 5: Rich world, poor world

> Hypothesis:
> 'European countries are wealthier than African countries.'

Write one or more hypotheses to compare aspects of life in two different regions.

- Use appropriate data and techniques to test your hypotheses, planning and specifying your methods carefully.
- You may choose to start with the hypothesis above, using the data in the following tables.

Appendix 2

Gross domestic product (GDP) is a measure of the wealth of a country. These tables give the GDP, in dollars, for 2003 for countries in Europe and Africa.

Europe

Country	GDP
Luxembourg	$55 100
Norway	$37 700
San Marino	$34 600
Switzerland	$32 800
Denmark	$31 200
Iceland	$30 900
Austria	$30 000
Ireland	$29 800
Belgium	$29 000
Netherlands	$28 600
United Kingdom	$27 700
Germany	$27 600
France	$27 500
Finland	$27 300
Monaco	$27 000
Italy	$26 800
Sweden	$26 800
Liechtenstein	$25 000
Jersey	$24 800
Faroe Islands	$22 000
Spain	$22 000
Man, Isle of	$21 000
Guernsey	$20 000
Greenland	$20 000
Greece	$19 900

Country	GDP
Slovenia	$18 300
Portugal	$18 000
Malta	$17 700
Gibraltar	$17 500
Cyprus	$16 000
Czech Republic	$15 700
Hungary	$13 900
Slovakia	$13 300
Estonia	$12 300
Lithuania	$11 200
Poland	$11 000
Croatia	$10 700
Latvia	$10 100
Bulgaria	$7 600
Romania	$6 900
Macedonia, The Former Yugoslav Republic of	$6 700
Turkey	$6 700
Bosnia and Herzegovina	$6 100
Cyprus	$5 600
Ukraine	$5 300
Albania	$4 500
Armenia	$3 900
Georgia	$2 500
Serbia and Montenegro	$2 300

Africa

Country	GDP
Reunion	$12 400
Mauritius	$11 400
South Africa	$10 700
Botswana	$8 800
French Guiana	$8 300
Seychelles	$7 800
Namibia	$7 100
Tunisia	$6 900
Libya	$6 400
Algeria	$5 900
Gabon	$5 500
Swaziland	$4 900
Guyana	$4 000
Morocco	$4 000
Egypt	$3 900
Lesotho	$3 000
Equatorial Guinea	$2 700
Ghana	$2 200
Guinea	$2 100
Angola	$1 900
Zimbabwe	$1 900
Sudan	$1 900
Cameroon	$1 800
Mauritania	$1 800
Gambia, The	$1 700
Senegal	$1 600
Togo	$1 500
Cape Verde	$1 400

Country	GDP
Uganda	$1 400
Cote d'Ivoire	$1 400
Rwanda	$1 300
Djibouti	$1 300
Chad	$1 200
Mozambique	$1 200
Central African Republic	$1 200
Benin	$1 100
Kenya	$1 000
Liberia	$1 000
Mali	$900
Guinea-Bissau	$900
Zambia	$800
Nigeria	$800
Madagascar	$800
Niger	$800
Ethiopia	$700
Eritrea	$700
Congo, Republic of the	$700
Burundi	$600
Congo, Democratic Republic of the	$600
Tanzania	$600
Malawi	$600
Sierra Leone	$500
Somalia	$500

[OCR, GCSE Mathematics, 1962/08/2345/2318/1969/05, 2005]

Task 6: Food for thought

'Cheese contains more fat than other foods.'

1. Use data (such as the nutritional information in the following tables) to decide how true this statement is.

2. Extend your investigation, making clear the rules and methods that you use.

All values in the nutritional content tables are based upon 100 g servings.

If a cell is empty there is no available value for the nutritional content. If the value is 0 then the food does not contain measurable traces of the nutrient concerned.

Cheese

		Cream cheese, low fat	Cream cheese, fat free	Brie cheese	Blue cheese	Camembert	Processed cheese, low fat	Creamed cottage cheese	Cottage cheese, low fat	Cream cheese	Edam	Feta	Goat cheese, soft	Parmesan, hard	Parmesan, grated	Ricotta cheese, full fat
Water	g	63.6	75.53	48.42	42.41	51.8	39.16	78.96	79.31	53.75	41.56	55.22	60.75	29.16	20.84	71.7
Energy	kcal	231	96	334	353	300	375	103	90	349	357	264	268	392	431	174
Total lipid (fat)	g	17.6	1.36	27.68	28.74	24.26	31.25	4.51	1.93	34.87	27.8	21.28	21.08	25.83	28.61	12.98
Sugars, total	g	0.2	0.4	0.45	0.5	0.46	0.51	0.29	0.33	0.2	1.43	4.09	0.89	0.8	0.9	0.27

		Gouda	Gruyere	Monterey Jack	Monterey Jack, low fat	Mozzarella	Mozzarella, skimmed milk	Processed cheese, full fat	Port Salut	Goat, hard	Cheddar, low fat	Cheddar	Cheshire	Neufchatel	Swiss cheese
Water	g	41.46	33.19	41.01	46	50.01	53.78	58.9	45.45	0.353	63.1	36.75	37.65	62.21	37.12
Energy	kcal	356	413	373	313	300	254	180	352	0	173	403	387	260	380
Total lipid (fat)	g	27.44	32.34	30.28	21.6	22.35	15.92	7	28.2	0	7	33.14	30.6	23.43	27.8
Sugars, total	g	2.22	0.36	0.5	0.56	1.03	1.13	0.59	0.57	0	0.52	0.52			1.32

180

Appendix 2

Meat, vegetarian meat and meat products

		Water	Energy	Total lipid (fat)	Sugars, total
		g	kcal	g	g
Pork tongue		56.9	271	18.6	
Luncheon meat		52.5	308	26.2	0
Frankfurter		57.82	278	24.31	0
Turkey, frozen, roast		67.84	155	5.78	0
Turkey breast		74.07	104	1.66	3.51
Chicken breast fried		60.21	187	4.71	0
Chicken, meat only fried		80.54	307	12.77	0
Chicken dark meat and skin, stewed		69.29	256	16.13	0
Chicken breast and skin, stewed		92.69	258	10.39	0
Chicken, meat only roasted		89.31	266	10.37	0
Chicken wings		13.40	94	6.32	
Chicken roasted with skin		47.87	288	15.32	

		Water	Energy	Total lipid (fat)	Sugars, total
		g	kcal	g	g
Canned chicken with broth		68.65	165	7.95	0
Canned turkey meat with broth		66.07	163	6.86	0
Non-meat sausages		50.4	257	18.16	0
Lamb casserole		55.77	269	18.05	
Roast shoulder of lamb		55.92	279	20.24	
Roast leg of lamb		60.82	231	13.69	
Beef casserole		66.19	160	5.52	0
Beef tenderloin roasted		57.55	211	10.32	
Pork salami		34.6	425	37	1.2
Roast ham		55.04	273	17.61	0
Pork shoulder roasted		54.8	292	21.39	0
Mortadella		52.3	311	25.39	0

If a cell is empty there is no available value for the nutritional content. If the value is 0 then the food does not contain measurable traces of the nutrient concerned.

Meat, vegetarian meat and meat products

		Corned beef, canned	Pork ham, canned roasted	Liver sausage (liverworst)	Grilled bacon	Bacon, pan fried	Non-meat bacon	Roast duck, no skin	Roast duck, with skin
Water	g	57.72	66.52	52.1	61.7	12.12	48.98	64.22	51.84
Energy	kcal	250	167	326	185	533	310	201	337
Total lipid (fat)	g	14.93	8.43	28.5	8.44	40.3	29.52	11.2	28.35
Sugars, total	g	0	0		0	0	0	0	0

Fruit and vegetables

		Carrots raw	Peas raw	Cauliflower	Green beans raw	Broad beans raw	Kidney beans raw	Mungo beans raw	Cabbage raw	Cabbage, savoy raw	Cabbage, red raw	Brussel sprouts raw	Spinach raw	Chicory (green) raw	Chicory (red) raw	Potato raw
Water	g	88.29	88.89	91.91	90.27	10.98	90.7	10.8	92.15	91	90.39	86	91.4	92	80	83.29
Energy	kcal	41	42	25	31	341	29	341	24	27	31	43	23	23	73	58
Total lipid (fat)	g	0.24	0.2	0.1	0.12	1.53	0.5	1.64	0.12	0.1	0.16	0.3	0.39	0.3	0.2	0.1
Sugars, total	g	4.54	4	2.4	1.4	5.7			3.58	2.27	3.91	2.2	0.42	0.7		

Appendix 2

		Water	Energy	Total lipid (fat)	Sugars, total
		g	kcal	g	g
Pickled sweet cucumber		78.51	83	0.57	1.51
Cucumber with peel raw		95.23	15	0.11	1.67
Cucumber peeled raw		96.73	12	0.16	1.38
Leek raw		90.5	34	0.4	
Spring onion raw		89.83	32	0.19	2.33
Red tomato raw		94.5	18	0.2	2.63
Cabbage lettuce raw		95.07	15	0.15	0.78
Cos lettuce raw		94.61	17	0.3	1.19
Iceburg lettuce raw		95.64	14	0.14	1.76
Sweet potato boiled and mashed		80.13	76	0.14	5.74
Sweet potato raw		77.28	86	0.05	4.18
Plain potato crisps		1.4	558	38.4	5.01
Baked potato no salt		47.31	198	0.1	1.4
French fried potato with salt		57.15	200	7.56	
Potato mashed with milk		78.51	83	0.57	1.51

		Water	Energy	Total lipid (fat)	Sugars, total
		g	kcal	g	g
Grapefruit pink raw		88.06	42	0.14	6.89
Grapefruit white raw		90.48	33	0.1	7.31
Grape raw with skin		80.54	69	0.16	15.48
Honeydew melon raw		89.82	36	0.14	8.12
Cantaloupe melon raw		90.15	34	0.19	7.86
Peaches raw		88.87	39	0.25	8.39
White onion raw		88.54	42	0.08	4.28
Apricots raw		86.35	48	0.39	9.24
Plums raw		87.23	46	0.28	9.92
Banana raw		74.91	89	0.33	12.23
Orange raw		86.75	47	0.12	9.35
Pear raw		83.71	58	0.12	9.8
Apple raw without skin		86.67	48	0.13	10.1
Apple raw with skin		85.56	52	0.17	10.39

If a cell is empty there is no available value for the nutritional content. If the value is 0 then the food does not contain measurable traces of the nutrient concerned.

Fast and convenience foods

		Cheeseburger with bun, plain	Burger with bun, plain	Pork frankfurter	Turkey frankfurter	Cheese spread	Frozen lasagne	Cheese-flavoured popcorn	Macaroni cheese canned	Quarter pound burger with cheese	Large burger (two burgers) with cheese	Baked potato with cheese sauce	Baked potato with cheese sauce and bacon	Muffin with egg, cheese and sausage	Cheese-flavoured potato sticks	Pop tarts cheese flavour
Water	g	38.65	42.1	59.85	62.99	58.5	81.39	2.5	81.39	48.36	50.54	65.75	65.02	47.4	1.8	16.85
Energy	kcal	329	311	269	226	295	82	526	82	269	261	160	151	295	496	407
Total lipid (fat)	g	17.83	16.73	23.68	17.7	28.6	2.46	33.2	2.46	14.4	14.1	9.71	8.66	18.7	27.2	17.7
Sugars, total	g			0	0	3.5	0.5		0.5							19.4

Appendix 2

		Water g	Energy kcal	Total lipid (fat) g	Sugars, total g
Baked potato with cheese sauce and bacon		65.02	151	8.66	
Baked potato with cheese sauce		65.75	160	9.71	
Onion rings breaded and fried		37.09	332	18.69	
Nachos with cheese, beans, beef and peppers		55.96	223	12.04	
Nachos with cheese and peppers		42.7	298	16.74	
Nachos with cheese		40.45	306	16.77	
Hot dog with bun and chilli		47.8	260	11.79	
Hot dog with bun, plain		53.96	247	14.84	
Chilli con carne		76.7	101	3.27	
Chicken fillet sandwich with cheese		46.01	277	17	
Chicken fillet with bun, plain		47.31	283	16.18	
Brownie		12.65	405	16.84	42
Tortilla chips with nacho cheese		1.7	320	3.93	1.92
Tortellini with cheese		30.5	307	7.23	0.95
Croissant with egg and cheese		45.47	290	19.45	

		Water g	Energy kcal	Total lipid (fat) g	Sugars, total g
Submarine sandwich with tuna salad		54.31	228	10.93	
Submarine sandwich with roast beef		58.97	190	6	
Mixed salad, no dressing with turkey, ham and cheese		82.45	82	4.93	
Mixed salad, no dressing with shrimp		89.1	45	1.05	
Mixed salad, no dressing with pasta and seafood		80.37	91	5	
Mixed salad, no dressing with chicken		87.05	48	1	
Mixed salad, no dressing with cheese and egg		90.46	47	2.67	
Mixed salad, no dressing		95.51	16	0.07	
Potatoes, hashed brown		60.14	210	12.8	
Mashed potato		79.21	83	1.21	
French fries in vegetable oil		35.35	342	18.43	0
Baked potato with sour cream and chives		69.45	130	7.39	
Baked potato with cheese sauce and chilli		70.08	122	5.53	
Baked potato with cheese sauce and broccoli		70.04	119	6.32	

185

If a cell is empty there is no available value for the nutritional content. If the value is 0 then the food does not contain measurable traces of the nutrient concerned.

Fast and convenience foods

		Egg scrambled	Egg fried	Egg hard boiled	Egg omelette, plain	Egg poached	Croissant with egg and cheese	Croissant with egg and cheese and bacon	Croissant with egg and cheese and ham	Croissant with egg and cheese and sausage	Pizza thin crust, cheese frozen	Pizza bread crust, cheese frozen	Pizza thin crust, pepperoni frozen
Water	g	66.7	69.13	74.62	75.83	75.54	45.47	43.92	51.14	45.87	46.28	43.46	42.44
Energy	kcal	212	201	155	153	147	290	320	312	327	268	260	296
Total lipid (fat)	g	16.18	15.31	10.61	12.02	9.9	19.45	21.98	22.09	23.85	12.28	8.78	15.2
Sugars, total	g	1.52	0.83	1.12	0.65	0.77							

Appendix 2

	Water (g)	Energy (kcal)	Total lipid (fat) (g)	Sugars, total (g)
Vanilla ice cream no added sugar	67.86	143	6.2	6.09
Vanilla ice cream fat free	64.35	138	0	23.88
Rich vanilla ice cream	57.2	249	16.2	20.65
Vanilla ice cream	61	201	11	21.22
Strawberry ice cream	60	192	8.4	4.43
Rich chocolate ice cream	56.55	255	16.98	19.81
Soft vanilla ice cream	59.8	222	13	21.16
Chocolate ice cream	55.7	216	11	25.36
French bread pizza with sausage and pepperoni	50.7	253	12.7	
Pizza supreme with sausage, bacon and onion	50.9	253	13.3	

[OCR, GCSE Mathematics, 1962/08/2345/2318/1969/05, 2006]

Index

A

Anscombe's quartet, 37–39
automation and statistics, 14–15
average, 72, 74

B

The best stats you've ever seen, 45
bias in sampling, 68
big data, 5
boxplots, 89, 91, 131
    using R, 132
Breathing Earth, 44

C

Cambridge Mathematics
    Framework, viii, 11
central tendency, measures of, 72–76
ChanceMaker, 92
collections, completing, 119–121
    worksheet, 122–123
computing, modern, 5
confirmatory approach, 33
Constituency Explorer, 24
context, importance of, 30
correlation and causation, 10
correlation coefficients, 35, 37

D

data, 10
data analysis, 25
data collection, 24–25
data presentation, 40–44
data visualisation
    dynamic, 44–46
    online, 24
    static, 40–44
A Day in the Life of Americans, 44
diagrams, ability to interpret, 7
distribution, 71, 131–133
    graphical representation and, 71–72
    and shapes, 9–10
    using ICT, 131
    using paper, 132
    using multilink cubes, 132
    worksheet, 133
dotplots, 76, 78, 128, 129
dynamic data visualization, 44

E

empirical distributions, 53–55
Excel
    adding a slider in, 75
    conditional formatting, 107
    random sampling, 101–103
expert statisticians, 1, 2
exploratory data analysis, 10, 32–35
    robust measures, 36
    teaching, 35–36
    using Gapminder, 113–115, 117–118
Exploratory Data Analysis, 32

F

Fathom, 87–91
finding fault, 106–107
    worksheet, 109–110
five-number summary, 34–39, 76, 131–132
formal inference, 92–93
functional statisticians, 1, 2–3
The future of data analysis, 32

G

GAISE pre-K–12 framework, 15–16
Gapminder, 44, 45
    exploratory data analysis, 113–115
    worksheet, 117–118
Gapminder Foundation, 45
Gnumeric, 98–101
graphical excellence, 40–41
growing samples, 66–67, 128–130
    worksheet, 128
Guidance for assessment and instruction in statistics education, 11

Index

H
heuristics, 24, 86
histograms, study of, viii–ix, 76
How faithful is Old Faithful, 25

I
inferential skills, 88
informal inference of data, 81–82, 139–140
    aspects, 82–83
    structuring statistical conclusions, 83–85
    and variability, 85–87
    visualisation software, 87–91
    worksheet, 142–144
interdecile range, 76–77
interpretation, aspects of, 26
interquartile range (IQR), 76

L
Large Hadron Collider, 14
law of large numbers, 52–53, 124–125
    worksheet, 125–126
Local and global thinking in statistical inference, 92

M
mathematics
    approach to learning, viii
    vs statistics, 5–6, 58
mean, finding 73–74
median value, finding, 74–75
meta-representational competence (MRC), 25–26
mid-range, 72
modelling, 55–56

N
New York Times, 64
NCTM data analysis and probability standards, 27–30

O
occasional statisticians, 1, 3

P
permutation simulation, 55
probability, ix, 5, 86, 92–93
    exploration of, 49
    pitfalls, 92
    proportion and, 139
    role in statistics, 26, 59, 65–66
progress measurement, 27
Pseudorandom numbers, 51
'the p-value problem,' 2

Q
quasirandom numbers, 51
questions
    formulating, 23–24
    finding good, 33

R
R, 95
    example activity, 97–98
    installation, 95–97
    useful commands, 144–153
random numbers, 51
'reading the data,' 9
repeated measurements, context of, 74–75
representations, 5–6
results interpretation, 26

S
sampling
    bias, 68
    populations and, 8–9
    target-error and population view, 68–69
    and variation, 58–59, 91, 128–133
scatter graphs, 7

signal and noise, 71
   graphical representation and distribution, 71–72
   measures of central tendency, 72–76
   measures of spread, 76–79
   shoe size activity, 134–138
simulations, 48–50
   completing collections, 119–121
   long-term stability vs short-term variation, 124–127
   practical considerations, 50
   worksheet, 122–123
spread, measures of, 76–79
standard deviation, 36, 37, 40, 59, 77
static data visualization, 40–44
statistical analysis, 14
statistical cycle, 17–23
   analysing data, 25
   collecting data, 24–25
   finding fault, 106–107
   formulating questions, 23–25
   interpreting results, 26
   measuring progress, 27–30
statistical enquiry (NCTM), 27–30
statistical inference, 84
statistical investigation, 15–17
   concept mapping, 18–19
   statistical literacy, 4, 6, 11, 17, 23, 55, 104
   understanding sampling and variation, 69
statistical reasoning, 83–85
statistical thinking, 58
statistics, vii, ix–x
   ability to interpret diagrams, 7
   correlation and causation, 10
   distribution and shapes, 9–10
   importance of context, 30
   new type of curriculum, 3–4
   recommendations vs practice, 5–6
   sampling and populations, 8–9
   student experience, 20–22
   teaching, 15, 22, 48, 71
   and technology, 14–15
   understanding variability, 7–8
   users of, 1–2
   vs mathematics, 5, 58
statistics education, 5–6

T

Teaching Probability, 26
Teach-Stat, 16–17
technology and statistics, 15–16
Thinking, fast and slow, 3
Tinkerplots, 87–91
total range, 76
Trendalyzer, 41, 45, 113, 117
Trilogy Meter, 43

V

variability, 7–8
   and inference, 85–87
variance, 37, 38
variation, 58–62
   key measures, 59–60
   due to measurement error, 77
   within distribution, 77–78
variability, types of
   error, 64
   external factors, 62–63
   measurement, 63–64
   natural, 65
   random, 65–67
   variables of interest, 64–65
The Visual Display of Quantitative Information, 40

W

Windy, 44